SpringerBriefs in History of Science and Technology

The *SpringerBriefs in the History of Science and Technology* series addresses, in the broadest sense, the history of man's empirical and theoretical understanding of Nature and Technology, and the processes and people involved in acquiring this understanding. The series provides a forum for shorter works that escape the traditional book model. SpringerBriefs are typically between 50 and 125 pages in length (max. ca. 50.000 words); between the limit of a journal review article and a conventional book.

Authored by science and technology historians and scientists across physics, chemistry, biology, medicine, mathematics, astronomy, technology and related disciplines, the volumes will comprise:

1. Accounts of the development of scientific ideas at any pertinent stage in history: from the earliest observations of Babylonian Astronomers, through the abstract and practical advances of Classical Antiquity, the scientific revolution of the Age of Reason, to the fast-moving progress seen in modern R&D;
2. Biographies, full or partial, of key thinkers and science and technology pioneers;
3. Historical documents such as letters, manuscripts, or reports, together with annotation and analysis;
4. Works addressing social aspects of science and technology history (the role of institutes and societies, the interaction of science and politics, historical and political epistemology);
5. Works in the emerging field of computational history.

The series is aimed at a wide audience of academic scientists and historians, but many of the volumes will also appeal to general readers interested in the evolution of scientific ideas, in the relation between science and technology, and in the role technology shaped our world.

All proposals will be considered.

Jens Høyrup

Explorations and False Trails

The Innovative Techniques That Brought About Modern Algebra

 Springer

Jens Høyrup
Section of Philosophy and Science Studies
Roskilde University
Roskilde, Denmark

ISSN 2211-4564   ISSN 2211-4572 (electronic)
SpringerBriefs in History of Science and Technology
ISBN 978-3-031-48157-4   ISBN 978-3-031-48158-1 (eBook)
https://doi.org/10.1007/978-3-031-48158-1

This Springer imprint is published by the registered company Springer Nature Switzerland AG
The registered company address is: Gewerbestrasse 11, 6330 Cham, Switzerland

If disposing of this product, please recycle the paper.

In memoriam Enrico Giusti
28 October 1940 to 26 March 2024

Preface

In July 2022, as I was standing at a bus stop in Trieste together with other participants in a conference, Matteo Valleriani invited me, if I had a book of not too large extent, to submit it to a book series of which he was an editor. At the time I had just finished the text of a more extensive book on "The World of the *Abbaco* – Abbacus mathematics analyzed and situated historically between Fibonacci and Stifel" [Høyrup 2024] and had in front of me half a year's work on preparation of that text for the publisher, and two other major impending tasks. At the moment I could therefore not take up the invitation – or gauntlet, or whatever you will call it.

However, during my work on the "World of the *Abbaco*" and spin-offs from that work I had become increasingly aware that its perspective was basically anthropological – it analyses each of the single historical stages on its own, evidently pointing out connections between them but deliberately staying away from the notorious "royal road to me", except for a few concluding pages. This seemed to call for a supplement, an analysis whose perspective was perpendicular to what I had already done – that is, an analysis showing *as such* the meanderings, hesitations, false trails and false and genuine starts that in the end produced, if not our present then at least the 17th-century beginnings of modern algebra. While algebra is an important topic in "The World of the *Abbaco*" it is definitely not the only one – from the anthropological perspective it could never be, abbacus algebra always thrived within a densely connected network, which also contributes to explaining the explorations, the hesitations, and the false trails. The complement instead had to concentrate on algebra. Other complements might of course have been, and remain possible: the career of the teaching of practical arithmetic until the 1950s at least; the on-off interest in the Euclidean theory of irrationals; or the occasional work on congruo-congruent numbers and other topics classified today under the theory of numbers – too discontinuous to constitute a "career". The present complement, in any case, deals with algebra.

I took up work in February 2023. As I anticipated, many of the objects in view were the same as in the previous book, just seen from a different angle; others had been hidden from view, and many of these I only became aware of during the work. In particular I now believe to understand how Viète and Descartes were pushed to introduce the use of abstract coefficients (mostly seen as the decisive step in the creation of modern algebra) and did not need to "invent" them. This is dealt with in Chapter VI. So is the (equally meandering) birth of the general parenthesis from Viète onwards.

The outcome is strictly a piece of local history (with a few remarks on the Arabic background), restricted to Latin and later Western Europe.

The outcome is here, with thanks for the invitation or gauntlet to Matteo. While "The World of the *Abbaco*" kept me occupied during the Covid-19 epidemic, forgetful of the work around me (not true, see [Booß-Bavnbek & Høyrup 2021]), the present work allowed me to make a soft landing.

Technically: All translations into English from original sources or secondary literature are mine where nothing different is specified. When translating, I try to keep as close to the original text as reasonably possible, often at the cost of stylistic elegance (with the exception due to type-setting convenience that fractions are mostly written with a slash, whereas the sources invariably use a horizontal fraction line). Terms and phrases in the original language as well as explanations from my hand may be inserted in square brackets. For ease of understanding I have added modern diacritics in words from medieval and Renaissance sources – thus *règle* and *lyée* where the original has *regle* and *lyee*. Since the present-day distinctions between *U* and *V* and between *I* and *J* were only established toward the end of the period I deal with, I have interpreted earlier spellings in agreement with present-day conventions – thus *unnd* instead of *vnnd*. Apart from that and from the expansion of abbreviations, quotations from original sources are intended to be exact.

Illustrations taken from manuscripts are redrawn for clarity, not reproduced directly.

References are made according to the author/editor-date system, in the format [NN year], or alternatively "NN ... [year]". The ordering of the bibliography is after first author.

Contents

Introduction

The changes that produced Modern algebra

Latin Europe first encountered algebra in the 12th century in the translation of al-Khwārizmī's algebra – that is, in this shape:[1]

> Divide ten in two segments and divide one of the two parts by the other, and four result. Whose rule is that you posit one of the two segments to be a *thing* and the other ten less a *thing*. Then divide ten less a *thing* by a *thing* so that four result. However, now you have known that when you multiply what comes out of a division by the same by which you divided, it will give back your amount which you divided. But what results from the division in this question was four, and that by which was divided was the *thing*. Multiply therefore four in a *thing*, and they will be four *things*. Thus four *things* are made equal to the amount which you divided, which is ten less a *thing*. Restore thus ten by a *thing*, and add the same to the four [*things*]. It will hence be that ten are made equal to five *things*. Now you have thus reduced this question to one of the six chapters, which is that roots are made equal to number.

Who is familiar with modern equation algebra only and never encountered rhetorical algebra will probably need to translate these lines into symbols in order to discover that they do indeed contain (equation) algebra.[2]

We may compare with some passages from Euler's *Introductio in analysin infinitorum*. First vol. I §139 [Euler 1748: I, 104–105]:

---

[1] From Gerard of Cremona's translation [ed. Hughes 1986: 248]. In order to make the structure of the argument as clear as possible I make this and later translations as literal as possible without loss of meaning or basic grammaticality.

For ease of understanding, I shall italicize *thing* throughout the book when it stands for an algebraic unknown. Similarly with its powers (presently *census*, later also *cube* etc.).

Gerard's translation was preceded by that of Robert of Chester, which however appears to have had a very restricted circulation. In practice, Gerard was the one who presented Latin Europe with the technique.

That Gerard's translation is much closer to al-Khwārizmī's original text than that of Robert, and also closer than the extant Arabic "living" text, is immaterial for the present discussion. Who is interested may have a look at [Høyrup 1998].

[2] The reader who feels the need may proceed thus: $10 = a+b$. Posit $a := t$, then $b = 10-t$, $(10-t) \div t = 4$, etc.

Let now in the formulas of § 133[3] n be an *infinitely small* number, or $n = \frac{1}{i}$, i being an *infinitely large* number. Then

$$\cos nz = \cos \frac{z}{i} = 1 \quad \text{and} \quad \sin nz = \sin \frac{z}{i} = \frac{z}{i}.$$

The sine of a vanishing arc $\frac{z}{i}$ is indeed equal to the arc itself, whereas the cosine $= 1$. Once that is posited one has

$$1 = \frac{(\cos z + \sqrt{-1}\ \sin z)^{\frac{1}{i}} + (\cos z - \sqrt{-1}\ \sin z)^{\frac{1}{i}}}{2}$$

and

$$\frac{z}{i} = \frac{(\cos z + \sqrt{-1}\ \sin z)^{\frac{1}{i}} - (\cos z - \sqrt{-1}\ \sin z)^{\frac{1}{i}}}{2\sqrt{-1}}$$

But taking the hyperbolic logarithm[4] above (§ 125) we have shown that $l(1+x) = i(1+x)^{\frac{1}{i}} - i$, or, writing y instead of $1+x$, that $y^{\frac{1}{i}} = \frac{1}{i} l y + 1$. If for y is written on one hand $\cos z + \sqrt{-1}\ \sin z$, on the other $\cos z - \sqrt{-1}\ \sin z$, then [from the first equation] it results that

$$1 = \frac{1 + \frac{1}{i} l(\cos z + \sqrt{-1}\ \sin z) + 1 + \frac{1}{i} l(\cos z - \sqrt{-1}\ \sin z)}{2} = 1,$$

since the logarithms disappear [because of the factor $\frac{1}{i}$ / JH], so that nothing follows. This first equation hence leads to nothing remarkable. However, the second equation about the sine yields:

$$\frac{z}{i} = \frac{1 + \frac{1}{i} l(\cos z + \sqrt{-1}\ \sin z) - \frac{1}{i} l(\cos z - \sqrt{-1}\ \sin z)}{2\sqrt{-1}},$$

whence

$$z = \frac{1}{2\sqrt{-1}} l \frac{\cos z + \sqrt{-1}\ \sin z}{\cos z - \sqrt{-1}\ \sin z},$$

from which follows how imaginary logarithms can be reduced to arcs.

Next the "general equation" for lines of the second order from vol. II §101 [Euler 1749: II, 48].

---

[3] [Respectively

$$\cos nz = \frac{(\cos z + \sqrt{-1}\ \sin z)^n + (\cos z - \sqrt{-1}\ \sin z)^n}{2}$$

and

$$\sin nz = \frac{(\cos z + \sqrt{-1}\ \sin z)^n - (\cos z - \sqrt{-1}\ \sin z)^n}{2}.$$

Here, n stands for any natural number./JH]

[4] [The "natural logarithm" of our terminology. Euler's "l" thus corresponds to our "log"./JH]

$$yy + \frac{(\varepsilon x + \gamma)}{\zeta} y + \frac{\delta x x - \beta x + \alpha}{\zeta} = 0$$

We need not go into the details of Euler's arguments, but we notice that now the presentation looks *grosso modo* as the equation algebra we know. We may feel unhappy with Euler's lack of rigour, finding perhaps a δ and an ε to be needed, but that is a different matter and concerns what happened after Euler's time to mathematical analysis, not what happened to algebra during the six centuries that separate Euler from Gerard's translation of al-Khwārizmī.

So, what has happened?

At the level of purpose we notice that al-Khwārizmī teaches a technique for solving problems. Euler is certainly able to do that too, but he also develops *theory*, and even when he solves problems he uses the theory he (or his predecessors) had developed.

What first strikes the eye, however, is the use of letter symbols; in Gerard's translation, as well as in al-Khwārizmī's original, everything including numerals is expressed in fully written words. If we go deeper into that question we observe that Euler's operations are performed directly at the level of these letters – which is precisely the reason we speak about them as *symbols* and not merely as abbreviations for words meant to serve within a verbal argument; verbal argumentation has certainly not disappeared, but it now surrounds the operations on the symbols, where it provides the framework of the logical structure.[5]

Introductory teaching of algebra in school at first presents symbolic algebra with a single unknown called x (that, at least, was the situation when I went to middle school, and I suppose that this is again how things stand in most of the world after the worst excesses of the new-math movement). That is close to al-Khwārizmī, the only difference being that he uses the word "*thing*". Elsewhere, al-Khwārizmī makes use of a *census*, which looks like another unknown; actually it is no independent entity but simply the outcome

---

[5] Cf. Nesselmann's often badly understood distinction [1842: 302] between three algebraic types – spoken of as *Stufen*, "stages", but clearly not meant by Nesselmann as being in necessary chronological order (then he should have believed Regiomontanus to precede Diophantos).

Nesselmann's "first and lowest" stage is that of "rhetorical algebra", in which everything in the calculation is explained in full words. The second, "syncopated algebra", makes use of standard abbreviations for certain recurrent concepts and operations, but "its exposition remains essentially rhetorical". The third is "symbolic algebra", in which

> all forms and operations that appear are represented in a fully developed language of signs that is completely independent of the oral exposition, thereby making every rhetorical exposition superfluous. We may execute an algebraic calculation from the beginning to the end in fully intelligible way without using one written word, and at least in simpler calculations we only now and then insert a conjunction between the formulas so as to spare the reader the labour of searching and reading back by indicating the connection between the formula and what precedes and what follows.

of the multiplication of the *thing* by itself.

Euler, on the other hand, operates with several genuine unknowns (two, x and y, in the equation from II.§101, elsewhere as many as he needs). Magnitudes that arise as powers of these, moreover, are now written is a way that excludes any idea that they might be unknown on their own. As regards the single unknowns, Euler designates their second power as a product, yy etc.; higher powers are denoted by exponents.

Modern school algebra, as also Euler but not al-Khwārizmī, also has special signs for mathematical operations: «+», «−», a fraction line standing for division, etc. ; and also an equation sign «=» (together with «<», «>», and many more). These signs can of course be spoken, but often in several ways – «+» thus as "plus", "and", "added to", etc.; moreover, they are not bound to a specific spoken language, they may be read in any tongue which serves to speak about algebra. They are *ideograms*, not *logograms* – signs referring to non-linguistic concepts and operations, not word signs.

As we see, Euler also has recourse to other letter symbols, α, β, γ, δ, ε, ζ. These stand where al-Khwārizmī and basic school algebra ascribe specific numerical values to the coefficients. The introduction of unspecified or "abstract" coefficients had as consequence that the *unknowns* become *variables*, though through a rather contorted process.[6]

All of these transformations are well known, and have often been pointed out. Less discussed, however, though also fundamental, it the appearance of the algebraic *parenthesis* – for instance in the equation

$$1 = \frac{(\cos z + \sqrt{-1}\ \sin z)^{\frac{1}{i}} + (\cos z - \sqrt{-1}\ \sin z)^{\frac{1}{i}}}{2}.$$

A *parenthesis*, it should be emphasized, is not a bracket, nor a pair of brackets. It is *what is enclosed by the brackets*, which can then be treated just as a single number would be treated – for instance, allowing Euler to raise $(\cos z + \sqrt{-1}\ \sin z)$ to the power $\frac{1}{i}$.

---

[6] From the semantics of the term one might suspect it to have resulted from the use of algebra or infinitesimal calculus to problems of movement, which turns out not to be true. Fermat, admittedly, uses the term occasionally in connection with imagined movement, but never in a context that can be characterized as algebraic. After 1650, the term turns up intermittently in more or less algebraic contexts without having the character of a technical term – thus in [Wallis 1655: 95] and in Johann Bernoulli's letters as reproduced in [Bousquet 1745: I]. Euler, on the other hand, treats it as a technical term when describing the first volume of the *Introductio* [1748: I, VIII] as dealing with "variable quantities and their functions" – this explicit function concept being in fact due to Johann Bernoulli [Youschkevitch 1976: 57–60]. Thereby we have *functions* of one or several variables. When such a function or set of functions are equalled to other functions or numbers we have one or more equations, which in a somewhat sloppy use of language become *equations* with one or more variables – "sloppy" because they are no longer variable but fixed (unless the equation or equation system is indeterminate).

That, however, happened well after the end of the period we are considering.

Euler, however, has other parentheses. The numerator

$$(\cos z + \sqrt{-1}\ \sin z)^{\frac{1}{7}} + (\cos z - \sqrt{-1}\ \sin z)^{\frac{1}{7}}$$

as a whole is indeed divided by 2. That is, the fraction line delimits the numerator as well as the denominator as parentheses. Even the root sign $\sqrt{\ }$ delimits the radicand as a parenthesis (as we shall see, these two types of parentheses precede the bracket-defined parentheses by centuries).

The present books aims at tracing these innovations, singly and in their interaction. It is thus an instance of "history in future-perfect", a question about how the present came about. But it is not a piece of retrospective teleology, explaining the historical process as resulting from a pull toward the present. If such a pull there was, it is not easy to discern it in the sources – nor do the actors seem to have been aware of it.[7] What we see is rather hesitating or even stumbling, in terms I have used on earlier occasions [Høyrup 2010; 2015]. We are in the same situation as Stendhal's Fabrice in *La chartreuse de Parme*, who arrives in the middle of the confusion of the battle of Waterloo – but only discovers afterwards that there *has been* a battle, and even a battle that decides succeeding history.

Algebraic background(s)

The process started with the translation of al-Khwārizmī's algebra, but not only. Some input – though much less influential than usually believed, only modestly effectual indeed from the later 15th century onward – came from Leonardo Fibonacci's *Liber abbaci* and *Pratica geometrie*. And then, which was the really important event, came the direct adoption of algebra into the Italian abbacus school tradition from the early 14th century onward.

Al-Khwārizmī

We shall take things in order, beginning with al-Khwārizmī. Around 820 CE, on the request of the Khalif al-Ma'mūn, he wrote the first extant – and quite likely the very first – *treatise* presenting what is important and beautiful in *al-jabr wa'l-muqabalah*.

The core of the art – the "six chapters" referred to in the initial quotation – are six equation types or "cases" (originally born as mathematical riddles) about a *māl*, a "possession" (appearing twice in this original sense, as "amount", in the quotation) and its square root; in Gerard's translation the *māl* becomes a *census*, in Italian *censo*.[8] In

---

[7] That is not to say that no such pulls existed, but then only in a quasi-Darwinian way, as mental "ecological niches". If a tool more or less accidentally produced by one worker turns out to be fruitful it may be adopted and further unfolded by others. *May*, need not – as we shall see clearly exemplified by the repeated introductions of several unknowns.

[8] This, beyond the rarity of manuscripts containing it, is one of the reasons we must regard as quasi-nil the influence of Robert's translation: he uses *substantia*, a perfect Latin translation which nobody adopted.

abbreviation (not symbols since they are not operated on here), C standing for *census*, r for *radix*, N for number and α for a coefficient revealed indirectly by the use of a plural:

Kh1	$C = \alpha r$ – first example $C = 5r$.
Kh2	$C = N$ – first example $C = 9$.
Kh3	$\alpha r = N$ – first example $r = 3$.
Kh4	$C + \alpha r = N$ – first example $C + 10r = 39$.
Kh5	$C + N = \alpha r$ – first example $C + 21 = 10r$.
Kh6	$\alpha r + N = C$ – first example $3r + 4 = C$.

Kh1–Kh3 are "simple", Kh4–Kh6 "composite" or "mixed". As we see, all but Kh3 are given in normalized form;[9] examples after the first one then show how to reduce non-normalized equations to normalized form.

Formally, all seem to treat the *census* as the basic unknown, and after finding the root all rules also state the *census*. Already al-Khwārizmī, however, reveals that he considers the *root* the genuine unknown. That can be seen in Kh3, the case to which the problem in our initial quotation is reduced; here, the normalized equation would *be* the solution (as indeed it is in the first example).

For all cases, first a rule is given. For Kh4 this one [ed. Hughes 1986: 234]:

> *Census* and roots that are made equal to number are as if you say, "a *census* and ten roots are made equal to thirty-nine dragmas", whose rule is that you halve the roots, which in this question are five. Then multiply these in themselves, and from them come twenty-five. To which add thirty-nine, and they will be sixty-four. Whose root you take, which is eight. Then diminish from them the half of the roots, which is five. Then there remain three, which is the root of the *census*. And the *census* is nine. And if two *census* or three or more or fewer are mentioned, similarly reduce them to one *census*. [...]

We know from Thābit ibn Qurrah [ed. trans. Luckey 1941] that this rule (and the corresponding rules for the other cases) came from the "*al-jabr* people", those whose practice al-Ma'mūn had asked al-Khwārizmī to put into a book.

Al-Khwārizmī, however, was not satisfied with that. He was a member of al-Ma'mūn's "House of Wisdom", and other scholars close to al-Ma'mūn's court were familiar with the lettered proofs of Greek geometry (both translating it and extending it independently), and we may guess that this was the reason that even al-Khwārizmī provided geometric proofs for the correctness of the rules. For Kh4 he even offers two, first this one – Figure 1 shows the accompanying diagram:[10]

> The reason, however, is like this. A *census* and ten roots are made equal to thirty-nine dragmas. Let thus there be made for it a square surface with unknown sides, which we

---

[9] This is different in the living Arabic text that has come down to us. The initial examples are the same, however. They are all normalized, showing that Gerard's text reflects the original.

[10] The upper diagram is the one given by Gerard [ed. Hughes 1986: 237], the lower diagram corresponds to what is found in the Arabic manuscripts [ed. Rashed 2007: 110*f*].

want to know together with its roots. Which is the surface *ab*. Each of its sides, however, is its root. And each of its sides, when it is multiplied by some number, then the number that results is the number of roots which are also as the root of this surface. Since it is said that with the *census* there are ten roots, I shall take the fourth of ten, which is two and a half. And I shall make of each fourth with one of the sides of the surface a surface. They thus make with the first surface, which is the surface *ab* four equal surfaces whose length is equal to the root of *ab* and whose width is two and a half. Which are the surfaces *g*, *h*, *t*, *k*.

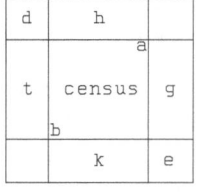
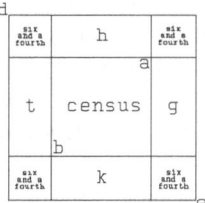

Figure 1

Joining these four rectangular surfaces to the first surface, and moreover the four small square lacking in the corners, each of area $2\frac{1}{2} \times 2\frac{1}{2} = 6\frac{1}{4}$ and together thus 25, gives us a larger quadratic surface equal to 64 and hence with side 8. Detracting in each end the fourth of the [number of] sides we are left with 3, which is the root of the *census*.

This argument proves that

$$r = \sqrt{39 + 4 \cdot \left(\frac{10}{4}\right)^2} - 2 \cdot \frac{10}{4} \ ,$$

not what is stated by the rule, namely

$$r = \sqrt{39 + \left(\frac{10}{2}\right)^2} - \frac{10}{2} \ ,$$

and al-Khwārizmī has to give additional arguments that the two are identical.

Next, however, al-Khwārizmī gives another proof. Stylistic arguments suggest that it has been added in a later revision of the text (but certainly by al-Khwārizmī himself) – see [Høyrup 1998]. Here, as shown in Figure 2, 5 times the side is added to two sides of the square representing the *census*. This is not Euclidean, but if al-Khwārizmī would make a proof based on *Elements* II.6 but explain it to an audience with no Euclidean training, this is exactly what he would do.[11]

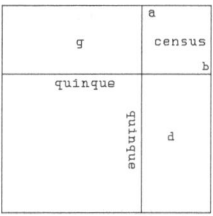

Figure 2

The first proof is definitely not Euclidean, but it corresponds to what is done in the Old Babylonian text BM 13901 #23 (from somewhere around 1800–1700 BCE), which finds the side of a square from "the four sides and the area" – a riddle which is also found in the pseudo-Heronic collection *Geometrica* 24.5 [ed. Heiberg 1912: 418] and in a number of Arabic mensuration treatises (and later in Fibonacci as well as Luca Pacioli). There can be little doubt that al-Khwārizmī knew it, and that he choose this symmetric but

[11] The orientation, we observe, is also the one found in the *Elements*. There are only four reasonable possibilities, however, of which this one is probably the one corresponding best to a right-left writing direction, so this could well be an accident.

somewhat inadequate way either because it was the first to come to his own mind or
because he expected it to speak more immediately to his readers.

These six cases with their rules were apparently seen by al-Khwārizmī as the core
(or the most "important and beautiful" part) of the discipline. Its name, however, "*al-jabr
and al-muqābalah*", was not taken from them. *Al-jabr* is the operation which we
encountered in the first quotation as "restoration", the reparation of a deficiency. There,
"ten less a *thing*" was restored as ten, and in order to preserve the equality, a *thing* also
had to be added to the other side.[12]

In the longer run, "restoration" was reinterpreted as the simultaneous addition of
something lacking to both sides of an equation, and *al-muqābalah*, "opposition", as the
corresponding subtraction on both sides. Originally, however, the meaning of the latter
term seems to have been the confrontation of two equal magnitudes as the sides of an
equation, perhaps in the production of the *reduced* equation; this would often involve
subtraction on both sides, which explains the reinterpretation (al-Khwārizmī's text is
ambiguous).

So, these operations (or at least restoration) are used, as we already saw in the
introductory quotation (above, p. 1). But al-Khwārizmī teaches even more – matters without
which is would be difficult to accept his book as an algebra in more than self-chosen name.
After the geometric proofs comes a "chapter on multiplication". It deals with the
multiplication of binomials, whose members may be added or subtracted numbers (including
fractions) or an algebraic *thing*. A "chapter of aggregation and diminution" then explains
the addition and subtraction of binomials involving numbers and surds or algebraic entities,
or even trinomials involving numbers, *census* and roots. But the chapter goes beyond its
heading and also deals with transformations of the types

$$a\sqrt{C} = \sqrt{a^2 \cdot C} \ , \quad \frac{\sqrt{p}}{\sqrt{q}} = \sqrt{\frac{p}{q}} \ , \quad \sqrt{p}\sqrt{q} = \sqrt{p \cdot q} \ , \quad a\sqrt{p} \cdot b\sqrt{q} = \sqrt{a^2 p \cdot b^2 q}$$

the last of which is derived from the first and the third.

For the addition and subtraction of binomials, al-Khwārizmī produces geometric
demonstrations, similar to what can be done in a two-dimensional coordinate system. For
the addition of trinomials, so he says, he had tried to make something similar, but it became
incomprehensible (as we can imagine), "but its necessity is clear in words".

Then follow first a chapter containing six problems, each reducible to one of the basic
cases – our initial quotation comes from the illustration of the third case; and then a
collection of miscellaneous problems, all reducible to one of the six cases. Gerard offers
12 problems, Robert somewhat more, and the extant Arabic text even more; some of the
latter are also listed by Gerard as having been found "in another book", an indication that

---

[12] Outside algebra, the term might also designate multiplicative restoration, for instance in Abū Bakr's
Liber mensurationum [ed. Busard 1968: 88], where $\frac{2}{5}$ of an area is restored through multiplication
by $2\frac{1}{2}$. For this operation, however, al-Khwārizmī uses other terms (in Gerard's Latin *reintegrare*).

he had access to at least two manuscripts but choose to translate the one which actually seems to be the best from a philological point of view.

Fibonacci

Fibonacci introduces algebra in chapter 15 part 3 of his *Liber abbaci*. A first version of this work was produced in 1202,[13] and a second version around 1228, at least not before 1226.[14] This part 3 at first [ed. Giusti 2020: 622–627] presents the foundations, and then [ed. Giusti 2020: 627–690] some 99 problems (the precise number depends on the extent to which we count variations as independent problems).

In the first section (probably going back to 1202), under the influence of al-Khwārizmī's geometric proofs, Fibonacci reinterprets the *census* as another term for square. His order of the cases is different from that of al-Khwārizmī, but there are echoes in his text that show him to have known Gerard's translation [Miura 1981] – yet echoes, no copying, indicating that Fibonacci was here engaged in an independent exposition.[15] That is confirmed by his geometric proofs, in particular those for Kh4. Like al-Khwārizmī he offers two but not the same. He has nothing similar to al-Khwārizmī's first proof. The second starts like this (the example still deals with a *census* with 10 *roots* being equal to 39 – see Figure 3):

Let there hereby be a tetragon *ABCD* having in each side more than 5 cubits, and let there be taken on the side *AB* the point *E* and on the side *AD* the point *F* and on the side *BG* the point *F* and on the side *CD* the point *H*, and let each of the straight lines *BE*, *CG* and *CF* and DF be 5 cubits, and let the straight lines *EH* and *FG* be connected. And because the quadrilateral *AC* is a tetragon, the side *DA* is equal to the side *BA*; and since, when from equals, equals are removed, what remains are equal, therefore when from *DA DF* is removed, and from *BA BE* is removed, each of which are 5, remains accordingly *EA* equal to the straight line *BE*. [...].

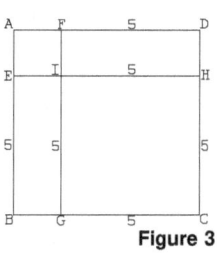

Figure 3

What follows after this piece of Euclidean rigour (the appeal to Common Notion 3, "If

---

[13] According to the Pisa calendar, we may assume, in which New Year was determined by Incarnation, meaning that the Julian date was between 25 March 1201 and 24 March 1202.

[14] For the problems of dating, see [Giusti 2020: xvii–xviii] and [Høyrup 2021b: 4, 18]. For convenience, and since the exact year is of no consequence for the present theme, I shall henceforth refer to "the 1228 version"

[15] At least in 1228 but probably also in 1202 (and certainly in the *Practica geometrie*), Fibonacci was faithful when copying while avoiding pastiche in his own supplementary explanations [Høyrup 2021b].

The echoes of al-Khwārizmī show, on the other hand, that here Fibonacci was inspired for his independent thought by a book, while the first fourteen chapters of the 1202 version appear to have been mainly based on oral interaction in the locations he had visited as a merchant.

equals be subtracted from equals ...") is not much different from al-Khwārizmī's second proof. But the latter is analytic, "treating the unknown as if it were known", while Fibonacci's procedure is rather synthetic ("having in each side more than 5 cubits"[16]).

Fibonacci is thus on his own. That is confirmed by his lettering of the diagram. A global survey of Fibonacci's writings shows that any copy from an Arabic or Greek original, or from a Latin translation from the Arabic or Greek which we know about, would use the alphabetic order of the source and speak of a square *ABGD*.

The second section contains several sequences of problems that can be show to be borrowed from Abū Kāmil though indirectly – one sequence at least from a treatise that was already translated into Latin (the Arabic original as well as the translation being now lost) – see [Høyrup 2022].

None of this – the ordering of cases, the synthetic approach, the problems indirectly borrowed from *Abū Kāmil* – are reflected in later sources before Pacioli (below, p. 11). Fibonacci was on his own. And stayed on his own. His fame was a product of late-18th- and 19th-century Italian attempts to create historical legitimacy for Italy as a nation-state. Well-deserved fame, certainly – he *was* a fine mathematician though hardly an "Italian mathematician" at a time where patriotism (coupled to mutual destructive enmity) was Genovese, Pisan or Florentine.

Fibonacci does not discuss the arithmetic of roots and binomials in his presentation of algebra. That is indeed the topic of his own extensive chapter 14, which goes far beyond what can be found in al-Khwārizmī, presenting also much material from *Elements* X though rather disorderly (almost certainly because much of this advanced stuff has been added in 1228 without thorough editing of what was there before).[17]

Abbacus algebra

The Italian abbacus school was a school for artisans' and merchants' sons, thriving between Genoa, Milan and Venice to the north and Umbria to the south. The first trace we have of it is a witness in a contract from Bologna from 1265, who characterizes himself as an abbacus master. In the 16th century it merged with the elementary school teaching reading and writing.

The abbacus school was mostly frequented by boys of age 11–12 for a year and a half or two years.[18] They were taught the Hindu-Arabic numerals and how to calculate

---

[16] As in Euclid, there is no explanation of where this "5" comes from (but Fibonacci evidently draws on the analysis he has found in al-Khwārizmī). Synthetic argument is the enemy of pedagogics (though perhaps a productive challenge for independent mathematical minds).

[17] See the analysis in [Høyrup 2021b: 18–25].

[18] Besides these there must also have been apprentices, future abbacus teachers. Two manuscripts also look as if they were made collectively by a master and his assistants/apprentices or the assistants alone (Vatican, Vat. lat. 10488, and Biblioteca Statale di Lucca, Codice 1754).

with them; in spite of what is sometimes believed because of the name it did not use any kind of reckoning board. It went on with the rule of three, metrological conversions and shortcuts; simple and composite interest; alloying; and perhaps the simple false position. Neither the double false position nor algebra were taught.

However, abbacus masters were members of a craft, and as such they were in competition – either for students, if they held private schools, as was mostly the case in larger cities, or for positions in the communal schools of smaller towns. For this purpose, they had to excel also in more advanced matters – and, as personal and professional identity goes, we may safely assume that they were also personally proud of this competence. Here, the double false position and algebra could serve.

We need not consider the double false position, since its interaction with algebra was utterly marginal.[19] Algebra was also more important, as confirmed by Pacioli.[20] In the ninth distinction of his *Summa* [1494: fol. 111^v] we read:

> I find that I shall no longer defer the part which is most necessary to the practice of arithmetic and also of geometry, in the vernacular commonly called "the major art" or "the art of the *thing*" or "*algebra* and *almucabala*", by us called "theoretical practice" [*pratica speculativa*]. Because in it are contained higher matters than in the minor art or mercantile practice.

Further on (fol. 144^r) we encounter another tribute to the algebraic art:

> Having with God's assistance come to the much desired place: that is, to the mother of all the cases by common people [*il vulgo*] called "the rule of the *thing*" or "the major art", that is, theoretical practice, also called *algebra et almucabala* in Arabic language, or according to some in Chaldeic, which in our language is as much as to say "of restoration and opposition, *algebra, id est restauratio, almucabala, id est oppositio vel contemptio, et solidatio*. Because in the said way infinite questions are solved. And those which still cannot be solved I shall point out.

When Pacioli wrote these eulogies, the discipline had developed for almost two centuries within the abbacus environment, and it was ready to confront the *Liber abbaci* – the Latin passage in the latter quotation draws directly or indirectly upon chapter 15 as well as chapter 14 of the *Liber abbaci*. As was to turn out, however, Fibonacci no longer had anything of interest to offer.

This part of the story had begun in the early 14th century, most likely in 1307. In

---

[19] An ultra-short description of its central principle is given in note 128.

[20] A similar eulogy is found on fol. 303^{3r} in the Ottoboniano *Praticha* (below, p. 21), where algebra is introduced:

> All that which has been said on this point would be in vain without the present, since here is shown the rule that solves all cases that can be solved, speaking in squares and cubes and in all the continuous quantities, as I shall show in the cases.

that year, a certain Jacopo da Firenze, working in Montpellier in Provence, wrote a *Tractatus algorismi*. Three manuscripts claim to contain it:

- Vatican, MS Vat. Lat. 4826, according to watermarks from around the mid-15th century;[21]
- Milan, Trivulziana MS 90, dated by watermarks to around 1410;
- Florence, Riccardiana MS 2236, written on vellum and therefore undatable.[22]

Only the Vatican manuscript contains an algebra. Together with a final collection of mixed problems this algebra might therefore seem to be a secondary insertion, as supposed by Van Egmond when he made his complete catalogue of abbacus manuscripts. However, stylistic analysis shows that the supposed secondary insertions share a number of characteristic stylistic features with the material that is shared by all three manuscripts, which makes it next to certain that the supposed insertions were already in the original (while the other two manuscripts represent an abbreviated version adapted to the school curriculum) – see [Høyrup 2007: 5–25].

However, even if the algebra in the Vatican manuscript should have crept in after 1307, other manuscripts show that this will have happened before 1327; for convenience I shall therefore speak of it as "Jacopo's algebra". Moreover, a number of other abbacus manuscripts from the earlier 14th century provide us with similar material though none of them so completely. There is therefore no doubt that "Jacopo's algebra" represents the new discipline, and it will be fitting to describe it as such (though with one necessary supplement).

The supplement belongs in the very beginning. Jacopo starts directly with the rules, "when the *things* are equal to number", etc. There is no introductory explanation, but we know from Jacopo's own words that something was lost, since the rules are followed by the remark "Here I end the six rules combined with various examples"; elsewhere such remarks invariably refer to what has been promised in the beginning of a section. The loss can well have happened during the transmission from Jacopo's original to the Vatican manuscript (which is at least two copyings removed from the original); less likely is that

---

[21] A dating "by watermark" of a manuscript means that some of its sheets carry watermarks that are also present in other documents that are firmly dated. In the actual case, two watermarks are found. One is also used in a document written in Pisa in 1440, the other in documents written in Siena between 1450 and 1452 and in Florence in 1453–54. See [Van Egmond 1980: 224] and [Briquet 1923: II, 344; III, 591]. As we see, this dating is not too precise. It is mostly better, however, than "internal evidence" when the date of a particular manuscript is asked for: a year of writing indicated in a colophon may well be copied from an earlier original (all manuscripts of Jacopo's *Tractatus* say 1307); the dates of real or fictional loan contracts are no better.

[22] Several scholars, including Warren Van Egmond [1980: 148] date it to 1307, but that is simply the year given in the colophon of all three manuscripts. Comparison with the Milan manuscript, to which it is close, shows the Florence manuscript to contain somewhat more errors and thus to be farther removed in the stemma (if not necessarily in time) from the original.

the introduction was already missing in an earlier treatise used by Jacopo (in that case, Jacopo would probably have produced a preamble of his own).

The beginning of the algebra of a certain Giovanni di Davizzo from 1339 is copied in a manuscript from 1424 (Vat. Lat. 10488. fols 29^v–32^r). This fragment seems to contain – not what was missing, since Giovanni's introduction is idiosyncratic (details later), but a parallel elucidating the kind of introductory material to expect: sign rules (plus times plus makes plus, plus times minus makes minus, etc.) and rules for the products of powers of the algebraic unknown and for operations with roots and perhaps binomials.

Returning to what we have – Jacopo's rules and the appurtenant examples – and comparing with al-Khwārizmī and Fibonacci – we observe some striking differences.

– All cases now deal with a *thing* (the *root* has disappeared). The *censo*[23] is still neither a primary nor some kind of parallel unknown; it is simply the product of the *thing* with itself.

– Probably in consequence of this shift the order of cases is now (*t* standing for the *thing*)

Ja1	$\alpha t = N$	Ja4	$\alpha C + \beta t = N$
Ja2	$\alpha C = N$	Ja5	$\beta t = \alpha C + N$
Ja3	$\alpha C = \beta t$	Ja6	$\alpha C = \beta t + N$

As we see, all cases are now presented in non-normalized form; accordingly, the first step in all rules is a normalization.

– There are no geometric demonstrations. The rules are simply stated as rules – the first runs:

> When the *things* are equal to the number, one shall divide the number in the *things*,[24] and that which results from it is number. And as much is worth the *thing*.

All six cases are provided with examples, sometimes one, sometimes two, the case Ja5 (the one allowing a double solution[25]) with three. Five are pure-number problems, five deal with pretended mercantile questions. Of the former, three are of the classical "divided 10" type, which we already encountered in al-Khwārizmī (above, p. 1). No examples are

---

[23] In order to keep in mind the character of the kind of algebra we deal with I shall use these loanwords (and the loan-translation *thing*) everywhere they are used in the sources, disregarding orthographic variation (e.g., *chubo* or *cubo*). When coming to German material, I shall use the German terms.

[24] The early abbacus texts distinguish division *in n*, referring to the division into *n* equal parts, and division *per n*, referring to the numerical operation. It seems that the abbacus authors did not understand the meaning of the underlying reasoning (they may "divide *per n* parts), but the normalization division (like, for instance, the division in partnership calculations) is invariably *in*.

[25] It should be observed that these two solutions (when they exist) are regarded as *possibilities* – if one does not work, the other certainly will. The unknown is really seen as an unknown but *already existing* number, and *not as a variable* that may take on different values fulfilling the given condition.

formulated simply in terms of *censi* and *things* (corresponding to al-Khwārizmī's *census-root-number*). The substitute (we may consider it a cheap way to create apparent complexity) consists of questions about numbers in given ratio. These were to become common in abbacus algebra; when more than two number are involved the ratios are always given so as to fit nicely together, for example as 2 : 3, 3 : 4, ... The numbers can then be posited as 2*things*, 3*things*, 4*things* (Jacopo's examples, being restricted to two numbers, do not demand this trick); the example for Ja2 shows how this allows to construct an example corresponding to a given case:[26]

> Find me two numbers that are in proportion[27] as is 2 of 3; and when each (of them) is multiplied by itself, and one multiplication is detracted from the other, 20 remains. I want to know which are these numbers.

Then there are the problems in genuine mercantile dress (numbers refer to case, letters to position among the examples for the case in question):

1b. There are three partners, who have gained 30 *libre*. The first partner put in 10 *libre*. The second put in 20 *libre*. The third put in so much that 15 *libre* of this gain was due to him. I want to know how much the third partner put in, and how much gain is due to (each) one of those two other partners.

4a. Someone lent to another one 100 *libre* at the term of 2 years, to make end of year.[28] And when it came to the end of the two years, then that one gave back to him *libre* 150. I want to know at which rate the *libra* was lent a month.

4b. There are two men that have *denari*. The first says to the second, if you gave me 14 of your *denari*, and I threw them together with mine, I should have 4 times as much as you. The second says to the first: if you gave me the root of your *denari*, I should have 30 *denari*. I want to know how much each man had.

5b. Somebody makes two voyages, and in the first voyage he gains 12. And in the second voyage he gains at that same rate as he did in the first. And when his voyages were completed, he found himself with 54, gains and capital together. I want to know with how much he set out.[29]

6. Somebody has 40 gold *fiorini* and changed them to *venetiani*. And then from those *venetiani*

---

[26] It also shows that Jacopo uses "restore" to designate subtractive (and elsewhere, as Fibonacci and al-Khwārizmī, additive) operations on both sides of an equation. *Opporre* (corresponding to Latin *opponere* and Arabic *muqābalah*) is absent from Jacopo's text; however, as mentioned above, p. 8, the original meaning of *muqābalah/oppositio* was probably the confrontation leading to the construction of a simplified equations, and this appears to be reflected in the term *raoguaglamento* used in example Ja5b [ed. Høyrup 2007: 316] about a simplified equation.

[27] Jacopo here and elsewhere writes *in propositione* – few of the early abbacus writers were familiar with proportion language (not to speak of proportion techniques and theory).

[28] That is, "to calculate and add interest at the end of year" – in other words, at compound interest.

[29] Both solutions are shown to be valid.

he grasped 60 and changed them back into *fiorini* at one *venetiano* more per *fiorino* than he changed them at first for me. And when he has changed thus, that one found that the *venetiani* which remained with him when he detracted 60, and the *fiorini* he got for the 60 *venetiani*, joined together made 100. I want to know how much was worth the *fiorino* in *venetiani*.

All deal with situations regularly treated in abbacus teaching, though in distorted ways – properly commercial questions were always of the first degree, second-degree problems could only be produced by asking odd questions. All also correspond to what is found in the later abbacus tradition. The "give-and-take" problem 4b is particularly noteworthy – the trick of introducing square roots or products in normally linear problems was to have a great future.

Within the calculations, Jacopo uses "restore" to designate subtractive as well additive operations on both sides of an equation. *Opporre*, as said, is absent from Jacopo's text; what we find is *raoguaglamento*, meaning "equation" or possibly "simplified equation". According to a fragment of another translation of al-Khwārizmī's algebra, made by Guglielmo de Lunis around the mid-13th century, this translates Arabic *elmelchel* (almost certainly reflecting *al-muqābalah* in Iberian pronunciation).[30]

At first one might believe these particularities to point to pre-al-Khwārizmīan algebra. So they probably do, but not directly. The pre-al-Khwārizmīan language is indeed still used by al-Karajī (later 10th and earlier 11th century). In the *Kāfī* [trans. Hochheim 1878: III, 10] he explains *al-jabr* to encompass additive as well as multiplicative completion (cf. above, note 12); in the *Fakhrī* [see Woepcke 1853: 64] he uses the same term not only for additive restoration but also for subtraction from both sides of an equation, exactly as does Jacopo. Moreover, he uses *al-muqābalah* about the formation of the reduced equation.

Al-Karajī's mathematics is much more advanced that what we find in early abbacus algebra. In the di-Davizzo fragment, however, we find matters that point to the innovations produced by al-Karajī. Firstly, di Davizzo, just like al-Karajī (see [Woepcke 1853: 53]), presents the sequence of algebraic powers and their reciprocals as two geometric series (although, as we shall see, di Davizzo goes wrong). Secondly, in the *Fakhrī* [Woepcke 1853: 57], al-Karajī formulates much better than Abū Kāmil [ed. trans. Rashed 2012: 316] the conditions under which sums of radicals can be simplified – for instance, that

$$\sqrt{8} + \sqrt{18} = \sqrt{2\sqrt{8 \cdot 18} + 8 + 18} = \sqrt{24 + 8 + 18} = \sqrt{50}$$

because 8×18 is a perfect square; exactly such transformations are also shown by Giovanni di Davizzo. All in all, abbacus algebra seems to have drawn much from an tradition that can be characterized as "diluted al-Karajī" (cf. [Høyrup 2011]).

---

[30] Two slightly different and partly independent versions of the fragment exist. Raffaele Canacci's *Ragionameni d'algebra* [ed. Procissi 1954: 302] speaks of *aghuaglamento* while Benedetto da Firenze [ed. Salomone 1982: 1] writes *asomigliamento*.

Not all of this is in Jacopo's presentation of algebra. However, after the examples for the first- and second-degree cases he offers rules (but no examples) for all but two of the cubics and quartics that can be reduced to first- and second-degree cases, or which can be solved by a root extraction (K stands for *cubo*, *CC* for *censo di censo*):

Ja7	$\alpha K = N$	Ja14	$\alpha CC = \beta t$
Ja8	$\alpha K = \beta t$	Ja15	$\alpha CC = \beta C$
Ja9	$\alpha K = \beta C$	Ja16	$\alpha CC = \beta K$
Ja10	$\alpha K + \beta C = \gamma t$	Ja17	$\alpha CC + \beta K = \gamma C$
Ja11	$\beta C = \alpha K + \gamma t$	Ja18	$\beta K = \alpha CC + \gamma C$
Ja12	$\alpha K = \beta C + \gamma t$	Ja19	$\alpha CC = \beta K + \gamma C$
Ja13	$\alpha CC = N$	Ja20	$\alpha CC + \beta C = N$

As we see (whether Jacopo saw it too we cannot know, since he just presents the rules), Ja11 (for example) can be reduced to Ja5 through division by t, and Ja20 reduces to Ja4 if we substitute C with t and in consequence CC with C. There is nothing similar in al-Khwārizmī's original text; in a sequence of borrowed problems late in Fibonacci's exposition of algebra something similar is presupposed [Høyrup 2022: 180–184], but we can be absolutely confident that this was not the inspiration – not least because Fibonacci does not take advantage of the reducibility but produces new geometric arguments.

The naming of the powers is to be taken note of. *thing*, *censo* and *cubo* might still look as distinct unknowns (although their mutual relation was of course known). Naming the fourth power *censo di censi*, on the other hand, shows the relation explicitly, in a way that excludes it could be overlooked by anybody within even a "diluted al-Karajī" tradition. Whether the composition was understood as a multiplication or as an "embedding" (to be explained below) cannot be seen directly, since $n^2 \cdot n^2 = (n^2)^2$; but since all higher powers were named by multiplication in Arabic algebra, as also in Italian writings for long, the multiplicative interpretation is not to be doubted.

After an oddly located alligation problem about grain follows in the Vatican manuscript a group of four problems which *we* would probably characterize as algebraic.[31] They deal with the successive wages of the manager of a *fondaco* (a warehouse located abroad – in Arabic, *funduq*), which are silently presupposed to grow geometrically. If the wages are a, b, c and d, the problems are

F1	$a+c = 20$, $b = 8$	F3	$a+d = 90$, $b+c = 60$
F2	$a = 15$, $d = 60$	F4	$a+c = 20$, $b+d = 30$

Jacopo offers nothing but unexplained rules and probably did not know how his rules had been derived; he has no recourse to algebra. None the less, it is rather obvious that the

---

[31] Wrongly, a purist might claim. Being algebraic depends on the way a problem is solved, not on the question. Even $3x+17 = 26$ is only algebraic to the extent we follow the invitation to manipulate the equation. If we use the double false position or trial-and-error it is not.

background is the kind of polynomial algebra that had been developed by al-Karajī.

Other abbacus writings from the next decades confirm this picture while providing it with more shades. First and most important there is a *Libro di ragioni* "Book of Problems" [ed. Arrighi 1987: 13–107], as Jacopo' *Tractatus* written in Montpellier but in 1327 and normally for brevity ascribed to Paolo Gherardi. The information found in the *incipit* is, however, that it is written "according to the rules and the abbacus course made by Paolo Gherardi of Florence", so it is plausibly written by a listener or an assistant. To this comes that the actual manuscript is a sometimes imprecise copy [Van Egmond 1978: 162].

Beyond some scattered problems solved by means of algebra, the *Libro di ragioni* contains in the end a systematic presentation of the field. Gherardi gives rules for these cases:

Gh1	$\alpha t = N$	Gh9	$\alpha K = \beta t$
Gh2	$\alpha C = N$	Gh10	$\alpha K = \beta C$
Gh3	$\alpha t = \beta C$	Gh11	$\alpha K = \beta C + \gamma t$
Gh4	$\alpha C + \beta t = N$	Gh12	$\alpha K = \beta t + N$
Gh5	$\beta t = \alpha C + N$	Gh13	$\alpha K = \beta C + N$
Gh6	$\alpha C = \beta t + N$	Gh14	$\alpha K = \beta t + \gamma C + N$
Gh7	$\alpha K = N$	Gh15	$\alpha K + \beta C = \gamma t$
Gh8	$\alpha K = \sqrt{N}$		

All rules are followed by examples. Some coincide with those of Jacopo, with or without the same numerical parameters; others are quite different.

Gherardi does not go beyond the third degree. On the other hand, he adds some new cases. Gh8 is solved correctly, $t = {}^{3}\!\sqrt{(\sqrt{N}/\alpha)}$[32] and may serve to remind us that surds did not belong to the number category. More stunning are Gh12, Gh13 and Gh14, for which false rules are given. Since the examples lead to solutions containing non-reducible radicals, a numerical test was not easily performed (in contrast to abbacus geometry, abbacus *algebra* never makes approximations). On the other hand, the rules given for Gh12 and Gh13 are identical, both copying that for Gh6. Anybody with insight in the matter would have seen that the solutions to Gh12 and Gh13 can only coincide if $t = C$, that is (zero being mostly not accepted as a number but only as a place-holder, and which in any case requires $N = 0$), $t = C = 1$ (which is only possible if $\alpha = \beta + N$). We must conclude that Gherardi either did not possess this insight or cheated.

Such false solutions survived in the abbacus environment for long, and their number would increase. They might serve in competitions for students or positions – neither municipal authorities not parents were likely to possess the qualifications necessary to expose them. However, they would not have been preserved if they had not served a

---

[32] To be precise, Gherardi gives the mistaken rule $t = \sqrt{({}^{3}\!\sqrt{N}/\alpha)}$, but the ensuing example is solved correctly.

purpose, that is, if the ability to solve higher algebraic problems had not been appreciated as something noteworthy (practical use can be excluded, both because any practical test would reveal the error and because there *was* no possible practical use). This created a pull, stimulating among those who looked through the fraud a conspicuous interest in doing better, producing in the 16th century the final break-through produced by Scipione del Ferro, Nicolò Tartaglia and Girolamo Cardano.

An anonymous *Trattato di tutta l'arte dell'abacho* was written in 1334, almost certainly in Avignon.[33] It exists in several copies, including the author's draft version (Florence, BNC, fond. prin. II,IX.57). The language is Tuscan, not Provençal, and the compiler is thus another Tuscan who had gone to Provence. It contains no systematic introduction to algebra,[34] but a number of problems are solved by means of *thing* and *censo*. They are similar to Gherardi's scattered problems and confirm that Jacopo, Gherardi and the present compiler drew on the same tradition. Since nothing similar is found in earlier Italian sources we must conclude that Jacopo and Gherardi learned their algebra in Provence, and that the "diluted al-Karajī tradition" on which they drew was thriving in the Ibero-Provençal area; we can exclude al-Andalus, Muslim southern Spain, as the place where they learned, since their texts contain no Arabisms).

The earliest extant abbacus manuscript produced within Italy and containing algebra is *Libro di molte ragioni d'abaco* from *ca* 1330 (Biblioteca Statale di Lucca, Codice 1754) (mentioned in note 18 as the outcome of a collaborative effort).[35] It contains no less that two presentations of algebra. Both depend, directly or indirectly but in any case strongly, either on Jacopo's *Tractatus* or on a close precursor of his while drawing to a limited extent on other material that had started to circulate (one new example corresponds, with changed numerical parameters, to a problem found in Gherardi).

This was thus the beginning of abbacus algebra, distinct from that of al-Khwārizmī though obviously related, and not at all inspired by Fibonacci. Its further broad development is described in [Høyrup 2024]. Here, we shall now turn to the single elements that in the end went into the new algebra of the 17th century. A short general account of the beginning of German *Coß* will also be needed, but it is better inserted in Chapter II when we approach German material for the first time.

---

[33] The date and probably place of writing were determined by Jean Cassinet [2001]. Reasons that the ascription to Paolo dell'abbacho must be rejected are presented in [Høyrup 2024: 215].

[34] After the draft manuscript follows the beginning of a systematic exposition, but it is in a different hand and of uncertain date. All it tells us is that it builds on the same basis as Jacopo and Gherardi.

[35] See, however, p. 46 below: Biagio "il vecchio" may well have worked on algebra before 1330. However, his kind of algebra, though independent of Jacopo, belongs to the same kind, and has the same roots.

Chapter I. Geometric proofs

Geometric proofs were given by al-Khwārizmī, as we have seen, and probably introduced by him. Fibonacci came to consider them the defining core of the algebraic art, as reflected in his reinterpretation of *census* as another term for square. They are absent from the abbacus algebra tradition – the very rare appearances have come in laterally.

Dardi da Pisa

The first intrusion is found in Dardi da Pisa's *Aliabraa argibra*, written in 1344, probably in Venice.[36] This is the earliest treatise we know about from the abbacus tradition which is dedicated exclusively to algebra; it is best known for dealing with no less than 194 "regular" and 4 "irregular" cases.[37] The former are cases for whose solution rules of general validity exist; all are given correctly by Dardi except two, where he has no terms for the fifth and the seventh root (see below, p. 26); heavy use of radicals allows Dardi to reach 194 cases. The first 16 cases do not involve radicals, but then these follow:

Da17	$N = \sqrt{(\alpha t)}$		Da22	$N = \sqrt{(\alpha K)}$
Da18	$\alpha t = \sqrt{N}$		Da23	$\alpha CC = \sqrt{N}$
Da19	$\alpha C = \sqrt{N}$		Da24	$N = \sqrt{(\alpha CC)}$
Da20	$N = \sqrt{(CC)}$		Da25	$\alpha t = \sqrt{(\beta t)}$
Da21	$\alpha K = \sqrt{N}$		Da26	$\alpha C = \sqrt{(\beta t)}$

Da21, as we see, coincides with Gherardi's Gh8; Dardi thus took initial inspiration from the existing tradition but expanded it immensely.

---

[36] Four manuscripts are known:
- Vatican, Chigi M.VIII.170 (*ca* 1395, cf. [Van Egmond 1980: 211]); in most though not all respects the best, and the one I shall use;
- a manuscript held by Arizona State University Tempe, written in Mantua in 1429; an essential complement to the Chigi manuscript;
- Siena, Biblioteca Comunale I.VII.17; *ca* 1470, ed. [Franci 2001];
- Florence, Biblioteca Mediceo-Laurenziana, Ash 1199, from c. 1495. I have only seen the extract in [Libri 1838: II, 349–356], according to which it is quite close to the Siena manuscript.

[37] Full lists in modern symbolism are given by Van Egmond [1983: 402–417] as well as Raffaella Franci [2001: 26–33]; Van Egmond also indicates the rule given by Dardi for each case in modern symbolic language.

© The Author(s), under exclusive license to Springer Nature Switzerland AG 2024
J. Høyrup, *Explorations and False Trails*, SpringerBriefs in History of Science and Technology, https://doi.org/10.1007/978-3-031-48158-1_1

The "irregular cases" are of the third and fourth degree and cannot be solved by simple root extraction or reduction to one of the six fundamental cases:

D-i1	$\gamma t + \beta C + \alpha K = N,$	D-i3	$\delta t + \gamma C + \alpha CC = N + \beta K$
D-i2	$\delta t + \gamma C + \beta K + \alpha CC = N$	D-i4	$\delta t + \alpha CC = N + \gamma C + \beta K$

Dardi offers rules that, as he says, are valid in particular situations only (situations which he does not specify). These rules can be seen to have been derived from homogeneous cases by a change of variable;[38] they were almost certainly not invented by Dardi [Høyrup 2024: 240].

Dardi's algebra, though going far beyond what he had inherited, thus belongs within the abbacus tradition. But occasionally he betrays familiarity with the Latin school tradition – both the way Latin was taught in grammar school and the university philosophers' idea of "four causes". He must also have seen Gerard's translation of al-Khwārizmī's algebra or a work descending from it, since he repeats his geometric demonstrations, adding one corresponding to the solution to Kh5 obtained by addition (that is, the solution 7 to the equation $C+21 = 10t$. Noteworthy is a change of style in the lettering. As we see in the diagram for the first mixed case, $C+10 = 39$ (Figure 4), equal areas are shown with the identical letterings. That appears to be Dardi's personal style (unless it is taken from an intermediate source which we do not know about.

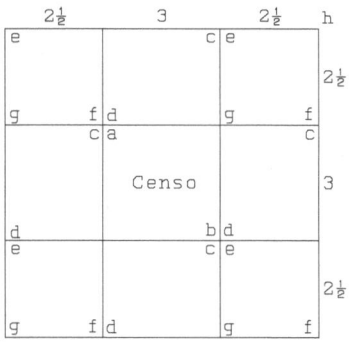

Figure 4

---

[38] Once we possess post-Cartesian symbolism this is easily done. For instance, the equation $x^3 = a$ has the solution $x = \sqrt[3]{a}$. If we replace x by $1+by$ we get the equation and solution

$$1+3by+3b^2y^2+b^3y^3 = a , \quad y = (\sqrt[3]{a}-1)/b.$$

Without this tool, it is an impressive feat. It is possible that the inventor of these solutions did not understand that they were not generally valid and that only Dardi realized this.

The Florentine encyclopedias

After Dardi we have to wait until *ca* 1460 before geometric proofs for the mixed algebraic cases turn up again.[39] It happens in three "abbacus encyclopedias":

- the anonymous *Libro di praticha d'aresmetricha*, Vatican, Ottobon. lat. 3307 (henceforth Ottoboniano *Praticha*);
- the equally anonymous *Trattato di praticha d'arismetricha*, Florence, BNC, Palatino 573 (henceforth "Palatino *Praticha*");
- Benedetto da Firenze's *Trattato di praticha d'arismetrica*, Siena, Biblioteca degl'Intronati, L.VI.47 (henceforth Benedetto's *Praticha*).

All three are autographs (several incomplete copies of Benedetto's treatise also exist). Benedetto's treatise is dated 1463, while the apparent dedicatee took possession of the Palatino *Praticha* in April 1460, which can then be taken to be the year is was written

---

[39] We might mention the appearance of geometric demonstrations in Giovanni Bianchini's Latin *Flores almagesti*, written from 1440 onward (I use Vatican, Vat.Lat.228, supported by Albrecht Heeffer's draft transcription [2015], to which he has gently given me access). Two early sections of this mostly astronomical treatise deal with arithmetic (fols 16^r–25^v) and algebra (fols 25^v–29^r, plus a problem collection left out in the Vatican manuscript but present in the transcription).

Bianchini, born in Bologna early in the 15th century, was active as a merchant in Venice after graduating from an arts faculty [Federici Vescovini 1968]. In Venice he seems to have frequented local abbacus masters (given his astronomical bent he will certainly have learned Euclidean mathematics at the university). In Venice he was discovered by Niccolò d'Este, Marquis of Ferrara, which led him to high posts in the finances of the Marquisate and also allowed him to dedicate time to theoretical as well as observational astronomy.

More than half of the arithmetic of the *Flores* is dedicated to the arithmetic of roots and arithmetical binomials. Here, a few geometric arguments are used (for instance, for the reduction of $\sqrt{24}+\sqrt{6}$); they are evidently not those used to demonstrate the rules for solving mixed algebraic cases.

The algebra as a whole seems according to the way the material is ordered to be an independent composition based on memory of what Bianchini had learned from his intercourse with Venetian abbacus masters and not a copy from some abbacus algebra (the same is probably true about the arithmetic). It is thus a witness of average early 15th-century abbacus algebra, but too distant from it to tell us anything with certainty; there will hence be no reason to mention it in later chapters.

Bianchini's algebra does offer some geometric proofs, inspired by those proposed by al-Khwārizmī. These he can hardly have encountered when frequenting the Venetian masters, but they are still so different (as regards numerical parameters as well as the lettering of diagrams) that he appears to have reconstructed them from memory (demonstrating thereby his mathematical competence) and not to have copied them from an earlier treatise. They are thus, it seems, not lateral intrusion into the abbacus tradition (as Dardi's proofs) but outside it. Interesting though they might be for a portrait of Bianchini, there is thus, even on this account, no reason to go in depth with them here.

(according to the introduction, however, it is a new version; fictional loan documents suggest that the first version was prepared between *ca* 1450 and 1454. The Ottoboniano *Praticha* refers to an event which Benedetto [ed. Pieraccini 1983: 118*f*] dates to 1445 as happening around 12 years ago. Combined with watermark evidence this suggests 1458 as its approximate date.

We know that these encyclopedias all come from a tradition encompassing Biagio "il vecchio",[40] Paolo dell'Abbacho, and Antonio de' Mazzinghi (active from the 1330s until the 1380s), whose works are mostly lost except for extensive quotations in the encyclopedias (we shall come back to Biagio and Antonio). It is not excluded that these predecessors had conserved memory of the geometric proofs. What is certain is that all wrote algebra in the abbacus tradition we know from Jacopo and Gherardi. They might still have shown al-Khwārizmī's proofs separately – that is what the encyclopedias do. However, this kind of Humanist philological piety belongs to the 15th rather than the 14th century. So, what is "not excluded" remains improbable.

Closer inspection of the three encyclopedias show that their source for the proofs is Guglielmo de Lunis's mid-13th-century translation of al-Khwārizmī (above, p. 15). Only Benedetto [ed. Salomone 1982: 1] refers to the Arabic *Aghabar* translated by Guglielmo, but terminological peculiarities (for example, the use of *recuperare* instead of *ristorare*) show that all three use the same source for this part of their exposition. The Palatino *Praticha* and Benedetto share a number of diagrams in this section, some of them not found in Gerard's translation, and the others with wholly different lettering. There is thus no doubt that the background is a different translation, and there is no reason to doubt the ascription to Guglielmo.[41]

---

[40] "The old" Biagio – old because there was also another, younger, abbacus master called Biagio.

[41] Raffaela Franci [2021: 231] claims that "an accurate reading of the text" reveals it to be that of Gerard.

The "accurate reading" consists of confrontation of two short passages from Gerard's translation with what is found in the Palatino *Praticha*; they are indeed fairly similar (one being in Latin and the other in Tuscan they are evidently not identical), *perhaps* somewhat more than would be expected from two independent conscientious translations of the same text though certainly not more that could happen if the Tuscan translator knew Gerard's Latin text (and medieval translators regularly took advantage of predecessors when producing improved versions). Franci has not compared with the Arabic text nor with any modern translation of that text, and may therefore find the similarity more striking than it is. Moreover, she does not observe that other parallel passages are definitely not close to each other. She notices the presence of some "complements" in the Tuscan text but overlooks that the letterings and orientations of shared diagrams differ, and also the use of *recuperare* instead of *ristorare*. Her observations could suggest that Guglielmo may have known Gerard's text, but does not prove even this hypothesis beyond reasonable doubt.

Pacioli

No later source I know of takes over Guglielmo's version of the proofs. Canacci, as mentioned in note 30, reports Guglielmo's Arabic-Tuscan equivalences, and he even has a couple of geometric proofs. These, however, are home-made – possibly inspired by Benedetto, whose treatise he may have known from his teacher Giovanni del Sodo (below, p. 38). Francesco Ghaligai, another former student of del Sodo, in [1521: 70^v] also takes over the list of equivalences with reference to Guglielmo and speaking of *rechuperatione*, but he has no geometric proofs at all.

Such proofs turn up instead in Pacioli's *Summa* in [1594], both in the arithmetical part and in the geometry. First the arithmetic (fol. 145^v and onward). Here it is obvious that Pacioli first takes over the *idea* of geometric proofs. He begins by proving the rule for *things* made equal to *censi*, which is of course easy but not to be found in any of his possible sources. After that the inspiration from the *Liber abbaci* is indubitable. Firstly, the composite cases are dealt with in the same characteristic order; secondly, the lettering of diagrams is the same. Thirdly, a first proof for the first composite case (as usually, $C+10t = 39$) follows Fibonacci's otherwise unique synthetic approach. In the end, however, Pacioli adds a reference to *Elements* II.4 as "proof of the proof". A second proof of the same case is similarly borrowed from Fibonacci, and here Pacioli adds the observation that *Elements* II.6 is used. Corresponding Euclidean references are added to the proofs of the other mixed cases.

The corresponding proofs in the geometric part (fol. 16^r onward) depend in a similar way on Fibonacci's *Pratica geometrie* [ed. Boncompagni 1862: 59]. Details can be omitted.

Cardano

It is worth taking note, instead, that with the exception of the comparatively uninfluential Ghaligai, 16th-century authors did not know Benedetto nor the other Florentine encyclopedias. What they knew was Pacioli, whose book had been printed, perhaps in 2000 copies [Sangster 2007], *all of which were sold* (in 1523 the same printer produced a second edition). Cardano thus knew about the possibility to provide geometric proofs for the basic algebraic cases from Pacioli. Certainly, like not a few contemporaries he loved to stand on Pacioli's shoulders while kicking his nose, so he would not confess the inspiration, and he would also change the numerical parameters – using as his first example in the *Ars magna* [1545: 9^v] $C+6t = 91$ – but a reference to *Elements* II seems to betray him. He has read more, however. While he says correctly that the proof in question differs from that of "Mahumete" (the name by which Gerard identifies al-Khwārizmī), an ensuing proof for the case $C = 6t+16$ takes over Gerard's diagram [ed. Hughes 1986: 240].[42]

---

[42] Cardano's very introduction also refers to "Mahomete" as well as Fibonacci, and on fol. 11^v he

These geometric proofs play no important direct role in the *Ars magna*. Cardano's own algebra, like that of his predecessors, is in the abbacus tradition albeit with some Euclidean condiments. However, when dealing with "the second unknown quantity multiplied", a new topic (fol. 23 onward), arguments become geometric again.

Indirectly, moreover, the geometric proofs not only entered: they may have been important in his solution of the third-degree equations. Once Tartaglia had divulged the solution to the equations $K+t = N$ and $K = t+N$, Cardano will immediately have recognized that the solutions were structurally similar to those expressing the sides of a rectangle whose area is given together with the sum of or the difference between the sides (and their relations to *Elements* II.5 and 6) – as expressed on fol. 16^r, "when, however, I understood that the rule which Tartaglia had divulged to me had been found by him by a geometric demonstration". Of course, Cardano cannot have known how Tartaglia discovered the solutions, but he here presupposes Tartaglia's way to have been the same as his own, which was thus exactly by geometric demonstration.[43]

In the distant-future-perfect perspective we may certainly see Pacioli's return to Fibonacci's geometric proofs as regressive – after all, abbacus algebra had developed to a point where proofs by means of polynomial algebra should have been possible. In the perspective of the closer future, however, things are turned upside-down. Even if we leave aside Leopold von Ranke's apparently more modest question about "how things really were" (as any serious historian will know, a no less difficult question and actually a haughty aim[44]) and just ask from the point of view of later times, we still need to decide *which* later time, and the answer depends on that choice.

---

states to have taken over two problems from the former.

[43] A hint (no more) that Tartaglia at least knew this geometric way is contained in a letter from 1539 reproduced in [Tartaglia 1546: 114^v–115^v]. Through a certain bookseller Zuanantonio, Cardano had asked for some of Tartaglia's problems with their particular solutions. Tartaglia refused, foreseeing that Cardano would then be able to reconstruct the rule.

[44] In von Ranke's words [1824: v*f*]:

> History has often been given the office to judge the past and to teach the present for the benefit of the future. The present essay does not take on so distinguished offices; it only intends to show *wie es eigentlich gewesen*.

That aim occupied the following 50 years of von Ranke's life. But still, as he says when starting this work in 1824, this first book of his (and all the others that followed), tells "nur Geschichten, nicht die Geschichte" – "only [partial] histories, not History".

Chapter II. Powers of the unknown

Geometry, as we shall see in Chapter VI, was to play an all-decisive role in the creation of the new algebra. However, the traditional geometric proofs – based on square-grid geometry and at times justified by references to *Elements* II – left the scene after having served Cardano.

So, let us turn to the properly algebraic topics – and first to the way to conceive and speak about the powers of the unknown. This will also allow us to introduce the algebraic writings that will be discussed under various perspectives in the following chapters.

"Powers", used in this sense, is strictly speaking foreign to most of the epoch we consider – it seems to have been introduced stepwise by Viète – more on this on p. 63. However, if we want to give a name to what we are doing when tracing the gradual emergence of the concept during a period where it did not yet exist and was not even sought for, there is no convenient way to eschew the seeming anachronism.

Abbacus algebra

Jacopo, as said above (p. 16), operates with *cosa*, *censo*, *cubo* and *censi di censi*, and his understanding of the last of these is certainly multiplicative. The same can be said about Gherardi, the *Trattato di tutta l'arte* and the *Libro di molti ragioni* (and many other abbacus writings from the later 14th and even the 15th century, which there will be no reason to mention).

The first evidence of conceptual expansion is found in the di-Davizzo fragment. As told on p. 13, it contains an introduction to algebra inserted in the manuscript Vat. Lat. 10488 from 1424. This latter manuscript is written in several hands, sometimes changing in the middle of a page, and sometimes (e.g., fol. 35^r) expressing personal opinions about its mathematics; we may conclude that it is the outcome of a planned collaborative effort, and that those who wrote it were mathematically competent. The best guess is that it was produced by an abbacus master and his assistants or apprentices, or by the latter alone.[45]

One thing we learn from the fragment is the names of powers:

---

[45] According to Van Egmond [1980: 230] it was written in Venice. He offers no evidence or arguments, and apart from an occasional use of *zenso* instead of *censo* (both in the same hand on fols 35^r and 38^r) and of *chaxa* instead of *casa* (both spellings in two consecutive lines on fol. 57^v) the language is Tuscan and certainly not Venetian. Since di Davizzo was from Florence and his manuscript likely to have been conserved there, Florence was probably the place where the manuscripts was composed. The inconsistent spellings may have resulted from use of a source written in a northern dialect.

J. Høyrup, *Explorations and False Trails*, SpringerBriefs in History of Science and Technology, https://doi.org/10.1007/978-3-031-48158-1_2

thing – censo – cubo – censo di censo – cubo di censo (or *censo di cubo*) – *cubo di cubo – ... – censo di censo di censo – ... – cubo di cubo di cubo di cubo.*[46]

This merely confirms that the naming was really multiplicative. More interesting is an attempt to give names to what we would speak of as negative powers, for instance, the outcome of a division of *censo di censo* by *cubo di cubo*. This outcome is claimed to be *root;* in general, what we might express as *thing^{-n}* is stated to be "*n*th root", with the exception that *thing^{-1}* is identified with *number*.

Di Davizzo evidently does not speak of an "*n*th root". Not only the notion of numbered "powers" was in the future, so were also the ideas of numerical exponents and numbered roots. His "roots" are composed multiplicatively from "roots" and "cube roots", similarly to the composition of positive powers from *censo* and *cubo*. Therefore, "dividing number by *cubo di cubo* gives cube root of cube root".

This would certainly not work if tried in calculational practice. Roots are by nature functions;[47] the cube root of 512 is 8, and therefore the cube root of the cube root of 512 is 2, the 9th root. But di Davizzo does not use his (substitutes for) negative powers in any kind of practice, his is a kind of experimental extrapolation not submitted to experiment.

Al-Karajī, in the *Fakhrī*, has a fully developed and operational understanding of negative powers, expressed as the corresponding "part" (*juz'*) – see [Woepcke 1853: 49]. Whether di Davizzo has known of any descendant of that and reconstructs badly or is fully on his own we cannot know. Later in the abbacus tradition, as we shall see, analogues of al-Karajī's notation turn up, but the idea is too close at hand to make this an argument for a borrowing.

Dardi da Pisa

Dardi reaches his impressive number of cases by means, not of higher powers but of radicals; his cases only speak of *censo* (spelled the Venetian way, as *çenso*), *cubo* and *censo di censo*. However, when deriving the rules he occasionally has to operates with higher powers. Thus, in #93, "number and root of *censi di censo* equal to *censi di censi*", the reduction gives rise to an eighth power of the *thing*, expressed multiplicatively as *censo di censo di censo di censo*. This is thus fully in agreement with the predecessors and uninteresting.

Interesting are instead his two erroneous solutions. In #177, a cube root is taken instead of a fifth root, and in #179 the root of the root is taken instead of the seventh root. As argued by Van Egmond [1983: 417], Dardi has no way to express these roots (Van Egmond

---

[46] These names are not given in an ordered list, they have to be read out from a sequence of rules, for which reason the names for the 7th, 9th, 10th and 11t powers are missing.

[47] Our concept, of course, going back to Johann Bernoulli, as we remember from note 6.

actually says "powers"): in contrast to di Davizzo he knows that the multiplicative terminology for the powers cannot be transferred to the roots.

Antonio de' Mazzinghi

After careful weighing of the available incongruous evidence, Elisabetta Ulivi [1996] reaches the conclusion that Antonio de' Mazzinghi was born between 1350 and 1355 and probably died between 1385 and 1386. He appears to have been a prolific mathematical writer, but most of his works including a *Gran trattato* have been lost; a copy of his *Fioretti* ("small flowers"), however, has been conserved in Benedetto da Firenze's monumental *Praticha* (on which much more later, in this as well as following chapters)[48]. It appears to be complete, and it is written with the express intention to "speak like master Antonio".[49] There is clear evidence in the text that what Benedetto renders is an unpolished work in progress – see [Høyrup 2024: 248].

Antonio was keenly interested in continued proportions and is supposed to have produced the first tables of compound interest. He also solves such problems as "to find the yearly interest equivalent to an interest of 20 percent made up every 9 months" and "to find the interest over 8 months that is equivalent to a yearly interest of 4 *denari* per *lira* and month" [ed. Arrighi 1967a: 35–30]. He thus knew extremely well how to operate with roots, and shows how to get around Dardi's stumbling-stone – namely by extracting in the second of these problems a fifth root called *radice relata*, (with the synonym *radice raportata* – both seem to mean "reported root"). Antonio does not claim paternity rights and may thus have taken over the terms from some predecessor unknown to us. On the other hand he points out that the problem had until then been considered impossible, which speaks in favour of a new introduction.

The innovation has a somewhat paradoxical impact on Antonio's naming of the algebraic powers: the Palatino *Praticha*, fol. 399^r [ed. Arrighi 2004/1967: 191] quotes Antonio for this explanation of the powers:

> *Thing* is here a hidden quantity; *censo* is the square of the said *thing*; *cubo* is the multiplication of the *thing* in the *censo*; *censo di censo* is the square of the *censo* [*quadrato del censo*], or the multiplication of the *thing* in the *cubo*. And observe that the terms of algebra are all in continued proportion; such as: *thing, censo, cubo, censo di censo, cubo relato, cubo di cubo*, etc.

The sixth power, as we see, is produced by multiplication; the fifth power could have been so too, as *censo di cubo* or *cubo di censo*. It appears that the new name for the fifth root has called forth a corresponding naming of the fifth power, but at the moment without general consequences being drawn.

---

[48] Fols 451^r–474^v, ed. [Arrighi 1967a]. Many of the problems are also conserved in the two other encyclopedias (see above, p. 21), but not with Benedetto's philological care.

[49] Fol. 460^r, ed. [Arrighi 1967a: 47]; cf. below, note 82.

The Florentine *Tratato sopra l'arte della arismetricha*

The manuscript Florence, BNC, Fondo princ. II.V.152, *Tratato sopra l'arte della arismeticha*, is an extensive abbacus treatise, a large part of which is dedicated to algebra; this part was edited in [Franci & Pancanti 1988]. According to watermarks and to fictitious loan contracts it was written in the 1390s by an anonymous Florentine [Franci & Pancanti 1988: 1]. Attempts to ascribe it to Antonio can be safely disregarded; their sole foundation is the assumption that there can have been only one eminent algebraist in Florence at the time. Antonio, however, was probably dead when this treatise was written, and many details of mathematical styles and terminologies of the two differ.

One of these differences has to do with the naming of algebraic powers. These begin with *thing, censo, cubo, censo di censo* – so far nothing remarkable. The fifth power is *cubo di censo*, a multiplicative composition. The sixth, in contrast, is *censo di cubo*, produced by embedding, as *censo(cubo)*, corresponding to a modern $(x^3)^2$, where *censo* is a function.[50]

So far this confirms what we might surmise from Antonio's inconsistent naming: that we are in the midst of a spontaneous conceptual phase transition, not really understood and certainly not systematically worked out by the participants.

But this *Tratato* offers more, "false trail" rather than mere "exploration". About the production of the second power it is said that

> Having seen what a *thing* means, having shown that it is a position, we come to its multiplication: we should know that a *thing* multiplied in itself makes a root which is called a *censo*, so that it is the same to say a *censo* as to say a quantity which has a root, engendered from a number multiplied by itself.

This is unobjectionable from a modern point of view. What follows is not. A *thing* multiplied by a *censo* gives

> *cubo*, that is a cube root, so that if you should say that if the *thing* should produce 6, then the *censo* will produce 36, that is, the square of the *thing*, the *cube* will produce 216 [...]. So it is the same to say *cubo* as to say a cube root of a given number.

---

[50] From a Piagetian point of view we may observe that it is probably easier to be familiar with the taking of the second power than with that of a third. As Piaget points out, it is easier to imagine picking a bouquet of roses than keeping together mentally a flock of swallows. In consequence, average children at the age of six say without hesitation that there must be more flowers than roses in the garden – while it may take years before they are equally convinced *a priori* that there must be more birds than swallows flying around. My experience with mathematics teaching, until engineering school and university freshman level in mathematics, confirms this to abundance. Once the independent variable is changed from t to the x they know from high school, the engineering students grasp immediately, and once the university students get a translation of a proportionality problem into a question about potatoes and money, they do too.

This conflation of power and root name goes on.

> *cubo di censo* [...] will be as much as saying a root which is engendered by a squared quantity multiplied against a cubed quantity; as it would be to say, if the *thing* were worth 6, the *censo* is worth 36, and the *cubo* will be worth 216, and the multiplication that is engendered by the 36 against the 216 will be 7776, you will thus say that if the *thing* were worth 6, then the *cubo di censo* will be worth 7776, and there are some that call this root the *radice relata*. So it is the same to say *cubo di censo* as to say *radice relata* of a quantity.

The author was a brilliant algebraist, as demonstrated by his handling of polynomial algebra [Høyrup 2024: 262*f*], and what to us looks as confusion did not disturb him; he understood to perfection what was in play, as shown by this passage:

> If you want to multiply a *thing* against a *cubo di censo*, it will be a *censo di cubo*, which means as much as to say, taken the root of some quantity, and of this quantity taken its cube root, as it would be if the *thing* were worth 3, the censo will be worth 9, the *cubo* will be worth 27, the *censo di censo* will be worth 81, the *cubo di censo* will be worth 243, the *censo di cubo* will be worth 729, because, taken the root of 729 it will be 27, whose cube root is 3, and that equals the value of the *thing*.

There is hardly any connection to di Davizzo's use of "roots" as negative powers. On the other hand it seems almost certain that the present "root names" for powers influenced the rest of the 15th century; in what way, however, is difficult to know. It is also doubtful that the present writer invented them. They may, like his reference to the "related root" as a term used by "some", point to something which the writer knows about from his environment.

The Florentine encyclopedias

As told above (pp. 21 and 22), the three encyclopedias all belong within a broad tradition going back to Biagio "il vecchio", and all three draw for the geometric proofs on Guglielmo de Lunis's translation of al-Khwārizmī's algebra. Close inspection of the texts show, however, that they have much more in common (the analysis can be found in [Høyrup 2024]). According to their own words, the compilers of the Ottoboniano and the Palatino *Pratiche* are both students of Domenico d'Agostino Cegia, a mathematical dilettante of standing and no abbacus teacher, generally known as *il Vaiaio*, "the fur dealer" – a reference to the profession of the family before protection by Lorenzo il Magnifico allowed it to improve its already substantial wealth and social standing [Ristori 1979; Ulivi 2002: 49*f*]. Their treatises can also be seen on the whole to be (sometimes free) redactions based on the same model – it is a plausible but unprovable guess that this model was put at their disposition by *il vaiaio*.[51] Benedetto can be seen also to have

---

[51] All three encyclopedias are indeed prestige objects dedicated to some protector-"friend" – the

known this model, or some very close kin, even though his own work goes far beyond it.[52]

All the more interesting is that the three do not deal with the algebraic powers in the same way (evidence that the situation remained in flux even in a closely knit Florentine environment around 1460.

The Ottoboniano *Praticha* (fol. 304[r–v]) offers this sequence (not as a sequence but within explanations of their products):[53]

> *dramme* (or number) – *thing* – *censo* – *cubo* – *censo di censo* – *censo di* cubo ("the same as *cubo di censo*") – *cubo di cubo*

"You may proceed with these multiplying infinitely". This is followed by divisions, first those leading to positive powers, as "dividing *censi* by *things* gives *things*", etc., then those leading to negative powers – for example (the coefficients 6 and 48 are used throughout in the examples illustrating divisions – those illustrating the multiplications use the corresponding 6 and 8),

> dividing *things* by *censi* gives a fraction denominated by *censo*, as dividing 48 *things* by 6 *censi* gives this fraction, that is, $\frac{8\ things}{1\ censo}$

Later (fol. 305[r]) it is pointed out that $\frac{8\ things}{1\ censo}$ can be reduced to $\frac{8\ dramme}{1\ thing}$.

The system for naming is thus purely multiplicative – there is nothing like the hesitations of the *Tratato*. The way to express negative powers is similar to that of al-Karajī, we observe (with the difference that al-Karajī makes use of the word "part" and not of the fraction line, which was only invented in the Maghreb in the 12th century).

What we find on fol. 372[r] in Benedetto's *Praticha* is clearly related: his examples for multiplications and divisions of powers use the same coefficients. But his powers are different:

> *thing* – *censo* – *cubo* – *censo di censo* – *cubo relato* – *cubo di cubo*

– that is, Antonio's system, as also quoted in the Palatino *praticha*. The *dramma* has

---

Palatino *Praticha* and that of Benedetto to members of the absolute top of the Florentine commercial elite (the dedicatee of the Ottoboniano *Praticha* is not identified). Putting such a model at the disposition of his students would thus have been a way for the *Vaiaio* to further their career.

[52] We are here looking at what they do in *their own* expositions of algebra (which is of abbacus type), not in the introduction they borrow from Guglielmo, which is there as a token of Humanist piety (as both Benedetto and the Palatino writer explain).

[53] *dramme* (*dragmas*) may be inspired by the Guglielmo extract – this use of a "unit" for pure numbers is rare in abbacus algebra but abounds in Guglielmo's text (more than in Gerard's translation).

disappeared,[54] and we find the same exception from the multiplicative formation as in Antonio. In the divisions, Benedetto differs from the Ottoboniano writer by presenting the result in reduced form immediately.

The Palatino *Praticha*, from which we know that this is Antonio's system (above, p. 27), also uses it, and similarly shows the outcome of divisions that lead to positive as well as well as negative powers. Its coefficients are different from those of Benedetto and the Ottoboniano writer (and changing); since the other two agree, the Palatino writer must be the one who is creative (on this modest level) in his relation to the model.

The Ottoboniano writer must also have been creative in his relation to the model used by all three – namely by discarding the inconsistent deviation from the multiplicative naming principle and returning to what he knows from the tradition.

An aside on the *regula recta*

In chapter 12 of the *Liber abbaci* [ed. Giusti 2020: 324], Fibonacci presents this give-and-take problem: A first man (A) asks from a second (B) 7 δ [*denarii*], saying that then he shall have five times as much as the second has. The second asks for 5 δ, and then he shall have seven times as much as the first. A first solution makes use of a line diagram. As an alternative, Fibonacci proposes a solution by means of the *regula recta*, the "direct rule", which "is used by the Arabs and is very praiseworthy". Here, B is posited to possess a *thing* (*res*) and 7 δ. After having received 7 δ, A therefore has 5 *things*, at first therefore 5 *things* less 7 δ. If instead B gets 5 δ from A, he shall have a *thing* and 12 δ, while A shall have 5 *things* less 12 δ. Therefore, a *thing* and 12 δ equals 7 times 5 *things* less 12 δ – etc.

To us, this is first-degree algebra with a single unknown. Fibonacci, however, does not conflate it with his notion of *al-jabr wa'l-muqābalah* (fundamentally a second-degree technique), and indeed introduces the latter technique only in chapter 15 part 3.

As discovered by Enrico Giusti [2017], a single manuscript of the *Liber abbaci* contains an early version of chapter 12, almost certainly the 1202 version. The present alternative solution is not in this early version of the chapter, but other references to the *regula recta* are – including some where the term covers first-degree algebra with several unknowns (more on this below, p. 78). Fibonacci appears at first to have taken over an Arabic method in 1202, and then when preparing the 1228 edition to have discovered the need for an explanation (this is documented in detail in [Høyrup 2024: 86–89]).

---

[54] In the beginning of the extract from Guglielmo (fol. 358^r), on the other hand, there is an interesting observations about the *dramma* – likely to have been added by Benedetto, since it is absent from the corresponding passage in the Palatino *Praticha* (fol. 391^r). Benedetto explains that "according to quantity, the *dramma* is tiny, as is the point with regard to the line". Benedetto appears to have derived a general dimensional understanding of the powers on the basis of al-Khwārizmī's text – but he wisely abstains from using it in his own algebra.

There is *a priori* no reason to assume that Fibonacci should dress up an invention of his own as if it were Arabic.[55] But we have more direct evidence for the correctness of his claim. In the *Liber augmenti et diminutionis* ([ed. Libri 1838: I, 304–371; ed. Moyon 2019b]) it is regularly used as an alternative to the double false position; in that work the name given to it is simply *regula* while the unknown is called *census* – translating Arabic *māl* and understood according to the original meaning "possession" or "amount of money".

As an aside to the aside we may take into account that in 1202 Fibonacci appears to have mainly presented material he had encountered in oral interaction as a Mediterranean merchant; that is told in his prologue, and confirmed by analysis of his text. *With the exception* of the algebra in chapter 15, however (cf. note 15 – detailed discussion with documentation in [Høyrup 2021b]). Merchants' mathematics, even when recreational and even when presented by teachers of future merchants, will normally be of the first degree.[56] That is, what Fibonacci presents as *regula recta* solution will have been shown to him by his commercial partners when they entertained each other with mathematical questions. They will not have been familiar with *al-jabr wa'l-muqābalah*, which was taught in the *madrasah* and not in the analogues of the abbacus schools; *that* technique Fibonacci therefore had to learn from books.

During the following two centuries, two known sources employ the technique. Firstly, the manuscript Paris, BN, Latin 15120, a modest 13th-century collection of mathematical problems, much of which is borrowed verbatim from the *Liber augmenti et diminutionis* – see [Sesiano 2000: 78–82]. Like its source it speaks simply of *regula*, but it changes the name of the unknown to *res*, "thing". Secondly, inspired by Fibonacci but sometimes going beyond him, Jean de Murs applies the method in the *Quadripartitum numerorum* under the name *ars rei*, "the art of the thing" [ed. L'Huillier 1990: 418, 420f, 436, 438, 456].

In known abbacus writings we find no references to the method before the three encyclopedias, all of which use it (as does Benedetto's *Tractato d'abbacho*, probably later than his *Pratica d'arismetricha*). They speak of it as *modo retto* and use *quantità* for the unknown (or for the primary unknown when several unknowns are involved). Striking is, however, a passage in the Ottoboniano *Pratica* (fol. 28ᵛ); the author refers to Fibonacci's understanding of the method and takes over his reference to its Arabic origin. But he also speaks about its use by "all the others who understand", and even points out that "some

---

[55] Not that such a manoeuvre is in itself unthinkable – we may consider Paracelsus's *alkahest*, certainly his own idea and no borrowing from Arabic alchemy in spite of its Arabic attire. A close analysis of Fibonacci's working methods, however, excludes it – see [Høyrup 2021b].

[56] *Normally* – there are exceptions. Most important of these are the riddles about a monetary possession and its square root which became the defining high point for *al-jabr*, cf. above, p. 5. The most sophisticated abbacus teachers also excelled in the transformation of originally linear recreational problem types into higher-degree questions by introducing products and square roots (above, p. 15). All of these, however, *are* exceptions.

say it is one of the exemplary modes of algebra". We shall more to say about this when considering the use of several unknowns (p. 89), and for the moment just make three observations.

Firstly, the method must have been sufficiently practised in earlier abbacus culture to have kindled discussions concerning its status.

Secondly (anticipating the following chapter), the *quantità* is regularly abbreviated *q* in marginal calculations. There was obviously no cognitive inability blocking the introduction of new symbolic abbreviations; the abbacus masters (at least those at the level of the writers we look at for the moment) were fully able to do that *when they found it adequate*. Writing for the audience of their own epoch and not for the historians or mathematicians of the 21st century, their choices do not always coincide with ours.

Thirdly – inviting a similar conclusion – in one linear problem, solved however by a procedure that leads to a reducible second-degree equation, the Ottoboniano writer (fol. 174^v in the margin, fol. 175^r in the running text) uses *quantità di quantità* about the square of the *quantità* (an emulation of *censo di censo*). This appears to be unique in the corpus as we know it: the *modo retto* is only used for linear questions. But again, once the need was there the writer – be it the Ottoboniano writer himself, be it a predecessor from whom he is copying – had no difficulty in inventing.

The end and aftermath of the abbacus tradition

In as far as algebra is concerned, the abbacus encyclopedias may be regarded as the high point of the abbacus tradition. After 1465 we encounter new experimentation but no establishment of a new consensus.

A Modena manuscript

The three encyclopedias may have been conservative in honour of the tradition to which they are outspokenly proud to belong. Materials from the next decades present us with new understandings, and they may well continue work which, unknown to us, was already going on in mid-century or even before.

Let us first look at the manuscript Modena, Bibl. Estense, ital. 578. It was probably written around 1485 ([Van Egmond 1980: 171], based on watermarks), and according to the orthography in the north – it uses *zonzi* and *mazore* where Tuscan would have *giongi* and *maggiore*. Van Egmond [1986: 23] observes that the manuscript is a copy. Part of his evidence is mistaken, but only part – see [Høyrup 2010: 41 n. 75]; the original thus precedes 1485, we do not know by how much. [Van Egmond 1986] contains an edition of the sections of the manuscript that deal with roots and with algebra, which however leaves out most of what is needed for the purpose of the present book; my presentation therefore refers to the manuscript.

Fol. 5^{r-v} lists the algebraic powers twice, first together with their appurtenant *gradi*, "degrees" (meaning steps upwards, for instance on a staircase), corresponding to our exponents. The naming is mixed, in part multiplicative, in part made by embedding (*gradi* in parenthesis here),[57]

number (0)

cossa (1)

zenso (2)

qubo (3)

zenso di zenso (4)

cossa di zenzo di zenso (5)

zenso di qubo (6)

cossa di zenso di qubo (7)

zenso di zenso di zenso (8)

cubo di qubo (9)

The second list (using the abbreviations introduced in the first list, here expanded) explains the "signification" in terms of root names:

cosa, its signification is that which you find

zenso, its signification is the root of that [which you find]

qubo, its signification is the cube root of that

zenso di zenso, its signification is the root of the root of that

cosa di zenso di zenso, its signification is its root of that

zenso di cubo, its signification is its root of the cube root of that

cosa di zenso di qubo, its signification is the 7th root of that

zenso di zenso di zenso, its signification is the root of the root of the root of that

qubo di qubo, its signification is the cube root of the cube root of that

Given the use of a 7th root, in the manuscript 7[a], it is a reasonable assumption that the grammatically problematic "is its" (*e la sua*) in the fifth line is a copying error for *e la* 5[a], "is the 5th, which in the ductus of the manuscript (and probably in that of the original) would look fairly similar. The equally problematic "its" in the next line could then be a set-off.

In any case we see that the writer does not know or at least does not employ the *relato* usage. We also see, from the reference to a 7th root but also by the way the 6th and the 9th root are expressed, that the writer is aware that roots cannot be produced multiplicatively – his roots are unambiguously *functions*, in spite of the less systematic expression of the powers.

As later Andreas Alexander and Christoph Rudolff (below, pp. 40 and 41, respectively), the present writer has a systematic exposition of how equations can be shown by means of reduction to be equivalent – first (in our usual abbreviation) the equivalence of $N+\beta t = \alpha C$ with $Nt+\beta C = \alpha K$, and so forth. For this, the namings and significations would have been of scant help, understanding would have asked for active thought. However, the *gradi*, though not spoken of here, will have been useful, allowing automatization (as we know from our own exponents).

In spite of sharing the idea of "root names" with the Florentine *Tratato*, the text certainly does not belong within its direct succession.

---

[57] Since the spellings determine the abbreviations that are used (to be discussed in the next chapter), I shall deviate from my usual principle and render the actual orthography.

Nicolas Chuquet and Étienne de la Roche

Certainly outside this succession, though part of the abbacus tradition, falls Nicolas Chuquet's *Triparty en la science des nombres*, written in Lyon in 1484 (Paris, Bibliothèque Nationale de France, Français 1346, ed. [Marre 1881]).

Chuquet [ed. Marre 1881: 299] presents himself as Parisian and *Bachelier en médecine*. He must thus have gone through first the Faculty of Arts and next that of Medicine there before settling in Lyon, the financial capital of France – presumably a profitable location for a teacher of practical arithmetic.

The manuscript goes well beyond what would serve a paying public, however. Chuquet has strong mathematical ambitions and opinions, and a level of mathematical competence justifying them. For now his naming by means of exponents is of interest. For roots, he uses ℞ provided with an ordinal written superscript to the right (fol. 45^v, ed. [Marre 1881: 103]). The *première racine* ("first root") of 12 is thus 12, and it is written ℞^{1}, "and so on for all the other numbers". The square root is thus ℞^{2}, the cube root ℞^{3}, etc.

Chuquet's terms for the sequence of algebraic powers (fol. 92^v, ed. [Marre 1881: 163] is

nombre – premier – second – tiers – ...

Accordingly, Chuquet speaks of algebra as *rigle des premiers*.

When used (fol. 95^r, ed. [Marre 1881: 165]), the powers are denoted by exponents – and nothing but. Instead of $12t$ Chuquet thus writes 12^1 (actually .12.^{1}, in agreement with the habit to set off numbers from the letters of the text by points, which makes the notation more transparent). 12^0 is simply the number 12, and multiplication of 2^1 by 2^1 gives 4^2. As we see, Chuquet draws full advantage of the notation.

A single manuscript of the *Triparty* has survived – not the autograph but a professional copy [Itard 1971: 272]. This copy was annotated by Étienne de la Roche and heavily used by him for his *Arismethique nouvellement composée* in [1520].

De la Roche took over the exponent notation for roots, but he did not possess Chuquet's systematic spirit. He drops the idea of a ℞^{1}, and writes the square root alternatingly as ℞ or ℞^{2}; for the cube root he proposes either ℞$^\square$ or ℞^{3}, and for the fourth root either ℞℞ or ℞^{4} [de la Roche 1520: 29^v]; only ℞^{5}, ℞^{6} and ℞^{7} are not provided with alternatives.

On fol. 42^r, de la Roche refers to Chuquet and to his *règle des premiers*, adding that "some nations call it *algebra* and some *almucabala*". De la Roche's own heading is *règle de la chose*, "rule of the thing". The exponent notation is shown on this page but afterwards dropped.

Since nobody else appears to have taken note of Chuquet's manuscript before Aristide Marre discovered it in the Bibliothèque Nationale, Chuquet's notation ended up having no influence in later times.

Pacioli and Pacioli

Thus was not the fate of Pacioli's *Summa de Arithmetica Geometria Proportioni et Proportionalita*, printed in [1494], with a print run that may have been around 2000 (above, p. 23).

However, before we consider this *magnum opus* we shall have a look at an earlier work from his hand, whose fate was not too different from that of the *Triparty*.

Pacioli was born between 1446 and 1448 in Borgo Sansepolcro (some 25 kilometres north-east of Arezzo) in a peasant family [di Teodoro 2014]. We know nothing about his mathematical upbringing, but he hardly went through an abbacus school. The suggestion that he was taught by Piero della Francesca, from the same city, remains a hypothesis. Around 1465, in any case, he moved to Venice, where he was the tutor of the sons of the merchant Antonio Rompiasi while also participating in the trading business; he also followed lectures at the Scuola di Rialto, at the level of an Arts Faculty, which indicates that he must already have acquired basic mathematical competence. In 1470–71 he was the guest of Leon Battista Alberti in Rome.

He was apparently never an ordinary abbacus master, but he taught abbacus-type mathematics repeatedly at the municipal *Studio* of Perugia, at a course connected rather to its Arts and Medicine than to its Law Faculty.[58] Here he produced a bulky manuscript, known from its incipit (and dedication) as *Suis carissimis disciplis*, "to his dear students".[59]

A systematic presentation of algebra has been lost from the manuscript, but we learn the names used for roots and powers in other parts of the manuscript. On fol. 313ʳ [ed. Calzoni & Gavazzoni 1996: 545*f*] we find the following roots (there is no complete list)

℞ – ℞ *cuba* (also written ℞*q*) – ℞℞ – ℞ *prima relata* – [...] – ℞ *seconda relata*

As we shall learn from the *Summa*, the *seconda relata* is the seventh root. The idea of multiplicative composition of roots has thus been left definitively behind.

At an earlier point (fol. 129ᵛ, ed. [Calzoni & Gavazzoni 1996: 186]), however, we find this sequence in the discussion of a problem about compound interest:

℞ – ℞ *cuba* – ℞℞ – ℞ *prima relata* – ℞ *cuba de* ℞ *cuba* – ℞ *de* ℞ *de* ℞ *cuba* –
℞ *de* ℞ *cuba de* ℞ *cuba* – ℞ *cuba de* ℞ *cuba de* ℞ *cuba* – ℞ *de* ℞ *de* ℞ *cuba de* ℞ *cuba*

Apart from the ℞ *prima relata* a "perfect" (and thus nonsensical) multiplicative composition. In which place (if not in both) Pacioli copies without thinking is difficult to know.

---

[58] [Rashdall 1936: II, 40–43]. Ulivi [2017: 6] offers a list of teachers of this course from 1406 to 1483.

[59] Vatican, Vat.Lat. 3129. One need not be a graphologist to derive from the writing that Pacioli was a personality *sui generis* and to be grateful to Guiseppe Calzoni and Gianfranco Cavazzoni for their edition [1996].

Both systems appear in any case to have been around.

The names for the powers are hidden within problems and mostly abbreviated (see the next chapter). Only interspersed commentaries allow us to identify them:

cosa – censo – cubo – censo di censo – primo relato – censo di cubo – secondo relato

This system is not multiplicative but indubitably made by embedding, and thus corresponds to the second naming of roots (which also precedes it).

The *Summa* contains several lists indicating the names of powers and roots. The first, in the margin of fol. 67^v, is accompanied by the observations that *tante terre, tante usanze* "as many regions, so many usages", and *tot capita: tot sensus*, "as many heads, so many opinions". It indicates a new version of the root names, abbreviations (on which more in the next chapter) and full names for the powers:

℞ 1^a	n°	*numero*
℞ 2^a	co	*cosa*
℞ 3^a	ce	*censo*
℞ 4^a	cu	*cubo*
℞ 5^a	ce.ce	*censo de censo*
℞ 6^a	p° r°	*primo relato*
℞ 7^a	ce.cu	*censo de cubo e anche cubo de censo*
℞ 8^a	2°r°	*secundo relato*
℞ 9^a	ce.ce.ce	*censo de censo de censo*
℞ 10^a	cu.cu	*cubo de cubo*
℞ 11^a	ce.p°.r°	*censo de primo relato*
[...]	[...]	[...]
℞ 29^a	ce.ce.2°r°	*censo de censo de secundo relato*
℞ 30^a	[9°] r°	*nono relato*

As we see, the names are now created by embedding, with the consequence that all powers that cannot be composed in this way from lower powers now become numbered *relati*; it is the complete version of the system that can be dug out from *Suis carissimis* and can be presumed to have been present in the lost algebra presentation of that manuscript (with different abbreviations, however, and not necessarily with appurtenant root names).

The second list (fol. 143^{r-v}) is a scheme showing how the "30 *gradi* of the algebraic *caratteri*" are brought forth as products. It identifies these *gradi* (also called *dignità* ... "since it is customary to call the said *caratteri* thus") (using both names and roots names, and mixing into one of the columns the corresponding powers of 2. We learn from this that Pacioli's contemporaries had a general term for "powers", namely *dignità* – Pacioli using also "characters"; moreover, that Pacioli shares the notion of *gradi* with the Modena manuscript but counts "number" as the first of these steps (meaning that they are *not* our exponents but correspond to the numbering of the root names).

All in all a perfect illustration of the adage that "as many heads, so many opinions".

del Sodo

Further illustration is offered by Pacioli's older contemporary del Sodo. According to [Ulivi 2017] del Sodo was born around 1420. After having been engaged in commercial activities he appears to have taken up teaching in rather mature age – Benedetto does not mention him in 1463 in a list of abbacus teachers, even though del Sodo seems to have been somehow connected to the *vaiaio*. In 1480 he had an exchange with Pacioli, mentioned by the latter. He seems to have taught until 1503 or even 1505, and died between late 1510 and early 1511.

No writing coming directly from del Sodo's hand is known, but his idiosyncratic names and symbols for the powers have been transmitted by his former students (above, note 30 and following text). The presentation in Canacci's *Ragionamenti d'algebra* [ed. Procissi 1954: 432*f*] is marred by inconsistencies], while the way the system is reported in Ghaligai's *Summa de arithmetica* [1521: 71^{r-v}] is consistent and thus probably to be relied upon. The sequence of powers is

*numero – cosa – censo – cubo – censo di censo – relato – cubo di censo –
pronicho – censo di censo di censo – cubo di cubo – relato di censo – tromico –
cubo di censo di censo – dromico – pronico di censo – cubo di relato*

The composition is made by embedding, and thus analogous to Pacioli's system in the *Summa*. The unique naming of those powers that cannot be composed from lower powers can be taken as evidence that the definitive acceptance of naming by embedding in the later 15th century gave rise to free experimentation in attempts to solve the problems inherent in the new system. The singular glyphs[60] used as abbreviations (see next chapter) support this interpretation.

---

[60] Sometimes, historians of mathematics consider anything standing for, say, a *censo* and not identifiable as a letter abbreviation as a "mathematical symbol", thereby blurring what is important is symbolism. Let me be emphatic: If somebody uses **ς** for the *censo* but still intends it to be read in normal language (just as Dardi's *ç* meant to be read *censo*), then it is part of a syncopated text and no more a mathematical symbol than "&" meant to be read "and". For such non-letter and quasi-letter abbreviations, no better word seems to be available than "glyph" – it is a logogram for a word, or an ideogram for a concept.

∀ and ∃ are mathematical symbols, *not* because they are not proper letters but because of the syntax within which they are used. And they remain symbols even though they abbreviate "[for] All" and "[there] Exists".

Algebra in German land

Experimentation was soon left behind, however. In the *Sesta parte del general trattato*, Tartaglia [1560: 1^r] repeats Pacioli's names and numbering for the *dignità* (stopping halfway, as the higher powers are unimportant, and leaving aside the root names); in the unpaginated chapter 1 of his *Practica arithmetice et mensurandi singularis*, Cardano [1539] does the same (using the same names again in the *Ars magna*). Only Bombelli [1572: 204] would replace *cosa* by *tanto* and *censo* by *potenza*.

We may therefore shift attention to what happened when a number of Latin-trained mathematicians from the German area discovered what to them, given their background, was a *new* mathematical art, namely algebra.[61]

Eclectic beginnings

At first these pioneers grasped what they could find. Part of their inspiration came from northern Italy, as reflected in the orthography of their technical loanwords: *coß* coming from *cossa*, not *cosa*, and *zensus* reflecting *zenso*, not *censo*. But they also learned from Florentine material (Regiomontanus certainly did), and even from the Latin translations of al-Khwārizmī's algebra; the latter, however, had little impact.

At first, the outcome was eclectic, the material at hand being disparate and those engaged in the process not yet competent enough to straighten the discrepancies. Details can be skipped, but see [Høyrup 2024: 381–395].

The first step toward some kind of streamlining can be found in the "Latin algebra" contained in the manuscript Dresden, C 80 (fol. 350^r–364^v, ed. [Wappler 1887: 11–30]). True, even this is an eclectic conglomerate, but the constituents overlap or supplement rather than contradicting each other. We shall return to it and in the present connection only observe that it uses abbreviations for the powers systematically. The names that can be identified in one of the segments are

res – zensum – cubus – zensum zensorum

In 1486, Johannes Widmann proposed and held a series of lectures on algebra in Leipzig – private but announced at the university [Gärtner 2000: 6, 34*f*]. At the time, Widmann was the owner of the Dresden manuscript, and since the announcement refers to a set of 24 rules and the "Latin algebra" does contain such a set, we may assume that his lectures drew on that conglomerate. In any case, there are no indications that Widmann's lectures – whether just repetitions of what was in the "Latin algebra" or further development – had any influence.

---

[61] We have some names: Regiomontanus, Friedrich Amann, and a Dominican friar known as Aquinas. Others, anonyms, were involved in the adoption too; we cannot be sure that all of these had first been trained in the Latin standard curriculum (Boethius, Euclid/Campanus, algorism, etc.), but it remains a reasonable guess – there is no trace of algebra in writings left by ordinary *Rechenmeister*.

Andreas Alexander

The real inception of the specific German algebraic tradition – the *Coß* – is due to Andreas Alexander, one of the first specialized matematics lecturers in Leipzig (no promotion but meaning that he might teach mathematics but would not be allowed to proceed to other more lucrative fields – *mathematicus non est collega*, in a saying from the later German Humanist tradition[62]) In his *Mathemalogium* [1504: C ii] he says that √32 is surd "for reasons that are shown to you in the art of algebra", and further (fol. Cihe states that algebra descends partly from arithmetic, partly from geometry "as I have explained to you in its text and commentary"; his unpublished algebra manuscript was thus written no later than 1504. Irrationals are indeed dealt with extensively in the manuscript, which also fits the other copious references to algebra in the *Mathemalogium*)

This algebra was never published but discovered by Menso Folkerts in the manuscript Leipzig, Hs. 1696. A closely related German algebra *Initium Algebrae*, "Algebra's Introduction", was published by Maximilian Curtze in [1902: 435–600].[63] It survives in no less than four manuscript copies, and consists of relatively short Latin quotations followed by long explanations in the vernacular. The order of the material is not the same in the two treatises. *Initium Algebrae* further offers geometric demonstrations for the algebraic cases which are not in the Latin manuscript.[64] However, the agreements are sufficient to lead Hellmann (who is preparing an edition of Alexander's algebra) to the conclusion that they come from the same hand [Rüdiger, Gebhardt & Folkerts 2023: II, 34]. Since the *Initium Algebrae* has the shape of a commentary to a Latin treatise, we may conclude with fair certainty that Alexander made several versions of his Latin algebra; this helps to explain that his work, though existing in manuscript only (but thus manuscripts, in the plural) could become as influential as we can see it to have been.[65]

---

[62] See, for instance, [Lindemann 1904: 7]. It seems that Alexander experienced this disdain: according to a letter from Melanchthon's hand, he was so despised in the university that he left Leipzig and made his living in Meissen by writing song-books and selling ink (ed. Martin Hellmann in [Rüdiger, Gebhardt & Folkerts 2023: II, 32].

[63] Because of its opening words *Initii Algebrae Arabis* [...] *prologus feliciter incipit*, Curtze assumed its fictitious author to be one "Initius Algebra". As argued by Hellmann [2003: 88], the phrase should be understood "Begins happily the prologue to the introduction of Algebra the Arab".

[64] They *might* be inspired by one of the translations of al-Khwārizmī, but then Alexander has extrapolated, offering also a proof for the case *census* equals roots. Inspiration from al-Khwārizmī is supported by the use of the term *radix* instead of *res* or *coß*.

[65] Apart from possible influence on Heinrich Schreyber and certain inspiration of Rudolff, his Latin treatise was also used by Adam Ries for the unfinished revision of his *Coß* [Rüdiger, Gebhardt & Folkerts 2023: II, 37].

In both (fol. 42^r respectively [ed. Curtze 1902: 474]), the sequence of powers is listed as

> *dragma – radix – zensus – cubus – zensus de zenso – sursolidum – zensi cubus – bissursolidum – zensus zensui de zenso – cubus de cubo*

Composition is made by embedding, as we see, with the consequence that new Latin names for the fifth and seventh powered have been introduced. In Alexander's Latin manuscript, the corresponding powers of 2, 3 and 4 are listed, in *Initium Algebrae* only the powers of 2. Both explain the nature of the sequence as a continued proportion, which allows them to reduce the 24 equations to 8.

The sequence expands the one we find in the Latin Algebra of C 80, and also its sequence of abbreviations.

Heinrich Schreyber

The first to bring German algebra into print was Schreyber, publishing under the Latinized name Grammateus. His *Ayn new kunstlich Buech, welches gar gewiß und behend lernet nach der gemainen regel Detre, welschen Practic, regel falsi unn etlichen regeln Cosse* includes a chapter [Grammateus 1521: Gvi^r–Liv^v] with "regula falsi together with several rules of *coß*", speaking much more about the *Coß* than about the double false position.

Schreyber knows the names "*radix, census, cubus, census de cen.* etc., but after mentioning them in the beginning (fol. Fvii^v) he goes on with discussion of continued proportions (and with their geometric interpretation as far as it goes, which already Alexander had given). This leads to his own naming and abbreviations (fol. Giii^r):

> *N* – 1*a:* or *pri:* – 2*a:* or *se:* – 3*a:* or *ter:* – 4*a:* or *quart:* – 5*a:* or *quint:* – 6*a:* or *sex:* – etc.

That is, the powers are named simply by the exponents, similarly to what Chuquet had done, but so differently that inspiration can be excluded (even if we disregard the difficulty that Chuquet's idea was buried in a manuscript kept in far-away Lyon)

Christoph Rudolff and later *coß*

The book that came to define *coß* for the rest of the century and beyond was Rudolff's *Behend unnd hübsch Rechnung durch die kunstreichen Regeln Algebra, so gemeincklich die Coss genennt werden*, published in [1525].[66] Rudolff has learned much from Alexander – he operates with the same 8 cases (fol. 52^r onward), and pokes fun at those who insist on 24. He also uses Alexander's names and abbreviations for the powers of the unknown (fol. 24^v). However, while Alexander offers no examples (except some

---

[66] Unpaginated; I refer to the handwritten foliation in the specimen in Augsburg, Staats- und Stadtbibliothek, Math. 740.

historical anecdotes, for instance about a riddle given to Solomon, which he then asked Algebras to solve), Rudolff has a huge number (many more than Schreyber). As regards the names and understanding of the unknown and its powers there is nothing important to observe (except Rudolff's systematic use of a *second* unknown, to which we shall return in Chapter V).

The same can be said about later important Latin works in the cossic tradition: Michael Stifel's *Arithmetica integra* from [1544], Johann Scheubel's explanation of algebra in his edition of *Elements* I–VI [1550] and Christopher Clavius's *Algebra* from [1608], and also about Valentin Mennher's *Arithmetique seconde* printed in Antwerpen in [1556], in which Rudolff's *Coss* was transferred into French.

Jacques Peletier

Slightly more important is Jacques Peletier's *L'algèbre* from [1554]. There was interest in the topic in France – Scheubel's introduction was published separately in Paris in [1551], and apparently sold well, since it was reprinted the next year.

Peletier points to Stifel (p. 4) as a main inspiration, but he has apparently also read Pacioli, Cardano and Scheubel; others he knows about without having seen their works – Rudolff, Adam Ries (whose *Coß* was a manuscript, only to be published in 2023), Pedro Nuñez (whose algebra was also a manuscript only in 1554, but to be published in 1567).

Peletier refers to the powers as *nombres radicaux*, which might point to Pacioli's root names. His names (p. 8) are inspired by those used by Stifel and in the cossic tradition:[67]

racine – çanse – cube – çansiçanse – sursolide

His glyphs are also those used by the Germans, the stylized *z* for *zensus* however replaced by a stylized *ç*, and the term for the first power replaced by ℞ in agreement with *racine*.

On p. 8 he invents the term "exponent" (*exposant*) by pure accident. It happens in a scheme that aligns the arithmetical progression $1 - 2 - 3 - 4 - ...$ with the abbreviations for the powers (a third line in the scheme indicated the corresponding powers of 2). The numbers in the arithmetical progression are said to *exposer* the *nombres radicaux* and their glyphs, and afterwards spoken the of as "the exposing", *les exposants* (meaning something like "the identifiers").

More interesting is what Peletier *does* with these exponents (p. 10). He points out that the prime factorization of the exponent shows how the corresponding powers can be understood as compositions by embedding – giving as examples 24 and 100. Thereby arises the need for names and glyphs for prime powers, which are introduced on p. 11. The cossic glyph for the *sursolidum* is ß, so for the 7th, 11th, 15th and 17th power Peletier uses bß, cß, dß and eß. We may think of Pacioli's *secondo*, *terzo*, *quarto* and *quinto relato;* maybe Peletier did so too.

---

[67] Peletier points out that "some" (probably Pacioli) speak of the fifth power as *premier relat.*

Jean Borrel's introduction to algebra [Buteo 1559: 117*ff*] is elementary and uninteresting in this respect (it stays at three powers). Guillaume Gosselin [1577], like Borrel eager to root algebra in classical Antiquity (though otherwise inspired by Tartaglia – in [1578] he translated the first two volumes of the *General trattato*), takes over many of the names and abbreviations used by Wilhelm Xylander [1575] in his translation of Diophantos (details below, p. 62).

With Viète (and then Descartes), the notion of *the* (or *the primary*) algebraic unknown is left behind. The chapter therefore ends here.

Chapter III. Abbreviations, glyphs, symbols and symbolic calculation

So, let us return in earnest to *symbolism*, and to those abbreviations and glyphs that should not be conflated with symbolism. Yet this asks for another detour. Algebraic symbolism had indeed been developed in al-Andalus or the Maghreb in the late 12th century (there are reasons to ascribe it to Ibn al-Yāsamīn, †1204, who moved between the two areas).

The Maghreb symbolism

Fibonacci knows about the fraction line and the notation for ascending continued fractions,[68] also apparently created in the Maghreb or al-Andalus in the 12th century. The symbolic notation, on the other hand, seems to have arrived too late.

This symbolism was "discovered" (in the same sense as Columbus discovered the Americas) by Franz Woepcke in [1854] in al-Qalasādi's *Kašf* from 1448 CE. In recent decades is has been described in detail by Driss Lamrabet [1994] and by Mahdi Abdeljaouad [2005; 2011; 2011]. Abdeljaouad [2005: 20, 24] has documented initial use in the 12th century (whether already fully developed or not is unclear) and survival in Ottoman mathematics at least until the late 18th century; Lamrabet [1994: 239] refers to use in the Maghreb as late as 1894.

The symbolism makes use of stylized and transformed single-letter abbreviations (thus half abbreviations, half glyphs) for powers of the unknown written over the coefficient (corresponding to the *thing*, the *census* and the *cube* – higher powers being produced by multiplication), and moreover for subtraction, division, equality, and square root. Where it serves, the stylized letters can be prolonged, thereby serving as algebraic parentheses; in that way, the glyph for division becomes similar to a fraction line.

This fraction line allows the writing of "formal fractions", expressions which have the shape of fractions but whose numerators and denominators are polynomials, and on which operations are performed *as if* they were genuine fractions, irrespective of the syntax of language proper. Thereby, the glyph becomes a genuine symbol and the notation a symbolism in the sense explained on p. 3. The other glyphs, even when able to function as algebraic parentheses, take as arguments only simple or composite numbers, not algebraic expressions. They are therefore not full-blown functions.

---

[68] For instance, "four seventh, and one half of a seventh", written $\frac{1}{2}\frac{4}{7}$. Fibonacci uses the notation in all surviving works and goes until ten levels.

In some manuscripts, the whole symbolic calculation is written into the text or as marginalia. Other manuscripts render fragments of fuller calculations that have been written on a clay- or dustboard [Abdeljaouad 2005: 29].

We have no direct proof that this Maghreb notation was taken over in abbacus algebra, but what turns up there is sometimes quite similar – sometimes not (in the latter cases thus either independent or misunderstood).

The Italian beginnings

The first trace of symbolic thinking in abbacus algebra is mutilated. It is found in Gherardi's *Libro di ragioni* (above, p. 17), in the illustration to the case Gh6, the division of 100 by some quantity and then by 5 more, the sum of the two divisions being given – in *our* symbols

$$100 \div q + 100 \div (q+5) = 20 .$$

Without making it explicit, Gherardi solves the problem by means of formal fractions. In the full form of the calculation (as found in later texts), the "quantity" is posited to be a *thing*, and the quotients then written as fractions and "added as if they were fractions"

$$\frac{100}{thing} + \frac{100}{thing \; plus \; 5} = \frac{100(thing+5) \; plus \; 100 \cdot thing}{thing(thing+5)} = \frac{200 \; thing \; plus \; 5}{censo + 5thing} = 20 .$$

Without mentioning fractions, Gherardi goes through the appropriate operations; he also refers to the scheme for cross-multiplication that produces the numerator (forgotten in the copy but easily reconstructible):

$$\begin{matrix} 100 & & 1 \; thing \\ & \times & \\ 100 & & 1 \; thing \; plus \; 5 \end{matrix}$$

The calculation is performed in detail in later abbacus books, which cannot have derived it from Gherardi's obscure rudiment; Gherardi must have borrowed it from earlier writers who also inspired other abbacus authors.

We must conclude that formal operations were known in abbacus algebra already before 1327. There is no trace at the moment, however, of any use of abbreviations, which teaches us an important lesson: *Just as abbreviations or glyphs are not necessarily meant as symbols and used in symbolic reasoning, symbolic reasoning can be performed with full words representing the entities involved – but then they are used not within the syntax of normal language but within the proper syntax of symbolic calculation.* When matters get complicated, symbolic reasoning evidently becomes prohibitively cumbersome if not supported by simple symbols. In other words, the link between *symbolic syntax* and *symbolic lexicon* is first of all a practical matter, no transcendental necessity.

Biagio "il vecchio"

The link, however, was not far away in time. Among other extracts from eminent predecessors, Benedetto da Firenze includes in his *Praticha* (above, p. 21) a large problem collection taken from an earlier *praticha*, this one written by Biagio "il vecchio" [ed. Pieracchini 1983] (above, p. 22). According to Benedetto, Biagio died around 1340, meaning that he may have been Gherardi's contemporary. Benedetto is philologically conscious and conscientious, and we may therefore feel confident that when he renders Biagio's symbolism in a way that differs from his own he follows, if not necessarily Biagio's original then at least the manuscript in his possession (at one point he actually doubts the fidelity of his manuscript[69]).

Biagio uses formal fractions repeatedly, also for more complicated purposes than Gherardi and most other abbacus algebras. We may look at one example (fol. 403^{r-v}; [ed. Pieraccini 1983: 110f]):

> Somebody makes two travels. In the first travel he earns 8, in the second he loses at the same rate as he has earned in the first. And then he finds to have 12 *fiorini*. It is asked with how much he set out. Posit that he set out with one *thing*, at the first travel, and he earned 8, he will thus have a *thing* and 8. And with this he left the first travel and goes to the second, at which it is said that he loses at the same rate as he has earned. You will say thus: at the first travel he made from one *thing* a *thing* and 8. And if at the second travel he loses at the same rate, he must do the opposite, that is that you will say, if from one *thing* he earned 8, he earned $\frac{8}{1\rho}$ of a *thing*, that is, $\frac{8}{1\rho}$ of his capital, where you will take $\frac{8}{1\rho}$ of a *thing* and 8, which are 8 *things* and 84 divided in a *thing*, that is, this fraction $\frac{8\rho\ 64}{1\rho}$.[70] And this is that which he find to have done in the second travel, and we said that he finds 12. $\frac{8\rho\ 64}{1\rho}$ are thus equal to 12. And in order not to have fractions, multiply both sides [*parte*] by 1 *thing*, you will have 12 *things* to be equal to 8 *things* and 64, where you shall confront [*raguaglierai*[71]] the sides, removing from both sides 8 *things*, and we shall have that 4 *things* are equal to 6, where the *thing* is worth 16. And we made the position that he set out with a *thing*, he thus set out with 16.

The glyph ρ for the *thing* (obviously not the Greek letter, but fairly similar, sometimes tending toward φ) was to become quite common in the Florentine tradition and beyond, though not without competition – "as many heads, so many opinions", in Pacioli's words. As we see, Biagio only makes use of this notation in the formal fractions, not in the running text. It already serves symbolic transformations, however – we may compare Biagio's

---

[69] Apparently without reason [Høyrup 2024: 225], which should enhance our confidence that what Benedetto reports really goes back to Biagio.

[70] Addition is indicated by juxtaposition, in agreement with the understanding of addition as aggregation of the constituents. The idea of arithmetical operations as mappings from $\mathbb{Q} \times \mathbb{Q}$ to $\mathbb{Q}$ is evidently out of place when we discuss pre-19th century mathematics.

[71] We observe the use of the equivalent of *muqābalah* for the production of the *reduced* equation – cf. above, p. 8. That subtraction on both sides is not meant is clear on fol. 303^v [ed. Pieraccini 1983: 112f], where we find "you shall confront the sides, giving to both sides 2160 and removing from both sides 234 *things*.

"in order not to have fractions, multiply both sides by 1 *thing*" with al-Khwārizmī's acrobatics (above, p. 1) – indispensable within his purely rhetorical exposition:

> Then divide ten less a *thing* by a *thing* so that four result. However, now you have known that when you multiply what comes out of a division by the same by which you divided, it will give back your amount which you divided. But what results from the division in this question was four, and that by which was divided was the *thing*.

At times *censo* appears in the formal fractions in full writing – thus fol. 303^v [ed. Pieraccini 1983: 112], $\frac{1080\rho\ m\hat{e}\ 2160}{2\,censi\ m\hat{e}\ 6\rho}$ (*mê* stands for *meno*, "less"); at times it is abbreviated c^o – thus fol. 405^r [ed. Pieraccini 1983: 125], $\frac{360}{1c^o\ 3\rho}$. In both cases, the equation in which they occur are reduced through multiplication by the denominator without further ado (symbolic syntax can still function without a full symbolic lexicon.

The former fraction results from the addition of $\frac{360}{1\rho}$ and $\frac{360}{2\rho\ m\hat{e}\ 6}$. The text refers to a marginal diagram as explanation of how to do it. That diagram (badly rendered by Pieraccini) shows the cross-multiplication similar to what has been lost in the surviving copy of Gherardi's *Libro di ragioni* (above, p. 45). Though probably widening their use, Biagio did not invent the symbolic calculations.

A modest beginning – yet still a beginning – though obviously not going beyond what was apparently already traditional in the Maghreb at the time. So far the question of influence or borrowing is undecidable (but we shall encounter supplementary evidence).

Dardi and *Alcibra amuchabile*

Dardi's *Aliabraa argibra* from 1344 (above, p. 19) offers a contrast. Dardi also makes use of abbreviations for *thing* and *censo*, but they remain abbreviations; in Nesselmann's terminology (above, note 5), Dardi's text is *syncopated*. *Censo*, in agreement with his Venetian spelling *çenso*, is abbreviated *ç* (for ease of distinction I shall use *Ç*), while the *thing* (*cosa*) is *c*.

Dardi's use of the fraction notation illustrates the difference. Before discussing that, however, we should take note of a particular use of ordinary fractions that is found in many abbacus treatises. Speaking (for example) of three men having money, they may write "the $\frac{1}{3}$" when intending "the third [man]". That is, the denominator of the fraction is understood as a *denomination*. In the "Columbia algorism" (plausibly the very earliest surviving abbacus text albeit conserved in a 14th-century copy only, see [Høyrup 2024: 182*f*]) this principle is used also [ed. Vogel 1977: 65] in such notations as $\frac{1}{grana}\frac{1}{2}$ standing for "1 *grana* and $\frac{1}{2}$ [of a *grana*]" – that is, in an emulation of an ascending continued fraction where the first "denominator" is a metrological denomination.

This habit explains Dardi's notation for algebraic monomials. On fols. 15^r, 22^r, 46^v of the Vatican manuscript we find $\frac{1}{\varsigma}$ standing for "1 *censo*", $\frac{30}{c}$ for "30 *things*", $\frac{21}{n}$ for "21 (in) numbers", and $\frac{36}{cu}$ for 36 *cubes*; further, $\frac{5}{c}\frac{1}{3}$ (an emulation of an ascending continued fraction) for "5 *things* and $\frac{1}{3}$ [of a *thing*]".[72] Nothing but a compressed linguistic expression and no operatory symbolism however rudimentary is intended. That can be seen from the way "1 *censo of censo*" is expressed (fol. 47^r) – namely as $\frac{1}{\text{Ç}}$ *de* Ç. In any case, no symbolic operations are possible on these pseudo-fractions "as if they were fractions".

That is not to say that Dardi does not know symbolic operations. He does, but they are of a different kind, based on graphic schemes and serving in the multiplication of binomials. For instance (fol. 6^r), for $(3–\sqrt{5})\cdot(3–\sqrt{5})$

(**R** abbreviates *radice*, "root"). It must be observed, however, that these schemes are used for operations with arithmetical binomials only (including square roots of numbers or of other arithmetical binomials), not for algebraic binomials.

Alcibra amuchabile[73] is a composite treatise dedicated to algebra alone from *ca* 1365. In a first part, where the arithmetic of roots and binomials is taught, schemes similar to those of Dardi are made use of. Since they are not quite as developed, Dardi is hardly the source – which on the other hand means that even here Dardi took over and unfolded a pre-existing technique. A second part lists 24 algebraic cases with examples. As much as possible is taken over verbatim from Jacopo's algebra.[74] Other problems, with false solutions, are shared with Gherardi, but not verbatim. A writer who copies one source faithfully would hardly treat another source differently without reason, which means that Gherardi was not the immediate source – with the further implication that Gherardi was not the inventor of the false solutions, just as he was not the inventor of the rudimentary handling of formal fractions (above, p. 45).

The third part consists of 41 solved problems. Interesting here for our present purpose is a full explanation of the use of formal fractions. The context, once again, is this question:

> Somebody divides 100 in a quantity, and then he divides 100 in 5 more than at first, and these two results joined together made 20. I want to know in what 100 was divided at first and in what it was divided afterwards.

---

[72] The notation is not Dardi's own invention. There is a single unexplained $\frac{10}{cose}$ on fol. 159^r of the draft version of the *Trattato di tutta l'arte dell'abacho* (above, p. 18).

[73] Florence, Biblioteca Riccardiana ms. 2263, fols 24^r–50^v [ed. Simi 1994].

[74] In one place, Jacopo gives up the transformation of $4\cdot\sqrt{54}$ into $\sqrt{864}$ and leaves spaces in the text, demonstrating thus that he calculates on his own (unless he copied uncritically from a source with this character). The *Alcibra amuchabile* performs the calculation, leaving no doubt that Jacopo's text (or, unlikely, this hypothetical very close precursor) is its source. The algebra in the Vatican manuscript of Jacopo's *Tractatus* cannot descend from the *Alcibra amuchabile*.

Then comes the explanation:

> Posit that you divided 10 in a *thing*, 100 divided in a *thing* results. And then say that you divide 100 in 5 more than at first, you shall thus divide 100 in a *thing* and 5, 100 divided in a *thing* and 5 results. Now you have to join 100 divided in a *thing* with 100 divided in a *thing* and 5. Now I will show you something similar so that you may be well advised about this joining and I will say thus: I will join 24 divided by 4 with 24 divided by 6, which you see should make 10. Thus posit 24 divided by 4 in the way of a fraction, from which results $^{24}/_4$. Also similarly posit 24 divided by 6 in the way of a fraction. Now multiply in cross, that is 6 times 24, they make 144, and now multiply 4 times 24 which is above 6, they make 96, join with 144, they make 244. Now multiply that which is below the strokes, that is 4 times 6, they make 24. Now you should divide 240 by 24, from which 10 should result. I say that if I multiply 10 which should result from it against the divisor 24, it will make the multiplied, that is, 240,[75] and so it does precisely. Let us therefore return to our problem. Let us take 10 divided by a *thing* and therefore posit these two divisions as if it were a fraction, as you see it drawn hereby. And now multiply in cross, as you did before, that is, 100 times a *thing*, which makes 100 *things*. And now multiply the other way, that is, 100 times a *thing* and 5, they make 100 *things* and 500 numbers; join to 100 *things*, you have 200 *things* and 500 numbers more. Now multiply what you have below the strokes, one against the other, that is, a *thing* times a *thing* and 5 more, they make a *censo* and 5 *things* more. Now multiply the results, that is, 20 against a *censo* and 5 *things* more, they make 20 *censi* and 100 *things* more,
>
> $$\frac{100 \qquad\qquad 100}{\textit{per una cosa} \quad \textit{per una cosa e più } 5}$$
>
> which quantity is equal to 200 *things* and to 500 *numbers*. Now take from each side 100 *things*, you will have that 20 *censi* are equal to 200 *things* and to 500 *numbers*. Bring to one *censo*, that is, that you divide each *thing* by the *censi*, you will have that one *censo* is equal to 5 *things* and to 25 *numbers*. [...].

We still see an echo of al-Khwārizmī's explanation from p. 1 in the statement that "if I multiply 10 which should result from it, against the divisor 24, it will make the multiplied, that is, 240". However, with the explanation that is given, this reference to (what we might see as) the definition of division becomes superfluous. As always in the extension of mathematical domains, the default assumption is that the familiar rules still hold; if they do not (as in the case of quaternions), serious and difficult work has to be done.

As we remember (above, p. 47), Biagio had felt no need to explain his ways to deal with formal fractions. History is not linear – and even less the "history" we construct from accidentally surviving sources.

---

[75] Thus, correctly, the manuscript. Simi writes 24, apparently taking the small final zero for a spot of ink.

The *Liber restauracionis*

The Latin *Liber restauracionis* (thus called by Marc Moyon, who in [2019a] made a critical edition of the three manuscripts known so far[76]) does not really belong within the abbacus tradition, nor is it however wholly foreign to it – border regions are often porous. Beyond these, a vernacular translation was made around 1400.[77]

The Vatican manuscript was published by Boncompagni in [1851]. Trusting its misleading incipit, he identified it with Gerard's translation. As this was proved by Axel Anton Björnbo [1905: 239–241] to be wrong, it seemed a natural assumption to ascribe it to Guglielmo de Lunis,[78] who is stated in a few late-15th- and early-16th-century sources to have translated al-Khwārizmī's algebra "into our language"[79] – cf. above, p. 22. This language could be Latin as well as an Italian vernacular. This identification was proposed by Moritz Steinscheider in [1904: I, 80], and by many others after him. Unfortunately, the ascription is impossible, an instance of fitting square pegs into round holes (or misuse of Occam's razor as an arrogant claim that we already know everything worth knowing – I leave to the reader to decide which is to be preferred). All references to Guglielmo's text[80] contain a long list of equivalences of Arabic and Italian terms (cf. above, note 30), and there is no trace of that in the *Liber restauracionis*. Beyond that, the part of the Guglielmo version we know from the Palatino *Praticha* has nothing to do with the *Liber restauracionis*. The latter remains anonymous.

The *Liber restauracionis* is no translation of al-Khwārizmī's text, not even a free translation. As discussed by Moyon [2019a: 8] it is either a *redaction* composed directly in Latin or a Latin translation of a redaction composed in Arabic during the late 12th or the 13th century; in either case, the author will have worked in al-Andalus or elsewhere in the Iberian Peninsula.

---

[76] Vatican, Vat. lat. 4606, fols 72^r–77^r – mid- to later 14th century; Oxford, Bodleian, Lyell 52, fols 42^r–49^v – probably early 14th century; Florence, BNC, Conv. Soppr. 414, fols 60va–63vb – maybe 1304.

[77] Vatican, Urb. lat. 291, fols 34^r–42^r, cf. [Van Egmond 1980: 215f]. Apparently based on the Latin Vatican manuscript.

[78] Probably identical with the Guglielmo de Luna who was connected to the Studio of Naples and/or to the courts of Frederick II and his son Manfred; and who translated Ibn Rušd and other philosophical writings into Latin. See [Delle Donne 2007]. Cf. above, pp. 15 and 22.

[79] All familiar from the preceding pages: Benedetto da Firenze, *Praticha*, fol. 368^r [ed. Salomone 1982: 1]; Canacci, *Ragionamenti d'algebra* [ed. Procissi 1954: 302]; Francesco Ghaligai [1521: 71^v]. Originally, only Ghaligai's text was known.

[80] Disregarding a wrong ascription to him of a manuscript of Gerard's translation – see [Hughes 1986: 223].

The reason to take up this treatise is a section [ed. Moyon 2019a: 17–20] *Qualiter figurentur census, radices et dragme*, "How *census*, *roots* and dragmas are rendered", apparently inspired by certain features of the Maghreb notation. *Census*, *root* and *dragma* are represented by c̲, r̲ and d̲, respectively (*dragma*, being just the unit of pure number, is not always written, or alternatively appears as a mere _); coefficients are written above, not below as in the Maghreb notation. The operator for subtractivity is a dot below the abbreviation. In the vernacular translation, as in abbacus algebra, the *root* has been replaced by *cosa*, abbreviated c̲, while conveniently *census* has become *sensus*, which can be abbreviated s̲.

Neither the Latin nor the vernacular version use the notation outside this short presentation; it seems to be something included for the sake of completeness but not part of the algebraic working of the author (if he has any). It thus represents no *symbolism*, not even a rudimentary first step. The existence of four surviving manuscripts is evidence of some interest, but mainly in an environment oriented toward Latin mathematical learning. After *ca* 1400 this interest waned; the notation seems to have been adopted by nobody else (which would indeed have been rather amazing, given that it is not shown in practical use).

Late 14th- and 15th-century algebra

After a small century, one might perhaps expect abbacus algebra to have reached some sort of homogeneous maturity in the late 14th century. This can indeed be claimed concerning the basic, school-bound level of abbacus mathematics. Abbacus *algebra*, however, was cultivated actively by relatively few teachers and therefore was not pushed toward homogenization. Theirs was a manuscript culture with little systematic exchange of new insights – competition, mostly concerned with the ability to solve intriguing problems, would rather make innovators keep for themselves their good ideas for solving these problems. Many sources (most, certainly) have been lost, so there may have been more of a continuum than we can see from what has accidentally survived – but what little has survived demonstrates beyond doubt that there was no unity and limited continuity.

Late-14th-century Florence

Florence was a large city – the fifth-largest of 14th-century Europe [Brucker 1969: 51]. It was an industrial, trading and banking city. No wonder that it was home to a number of abbacus schools and abbacus masters.[81] Some of them (Biagio "il vecchio", Paolo dell'Abbacho and Antonio de' Mazzinghi, cf. above, p. 22) seem to constitute a "tradition". Those who were active at the same time will evidently have known about each other as colleagues and/or competitors. To which extent they will have known about the *mathematics*

---

[81] See [Ulivi 2004] and [Ulivi 2002: 195–209].

of these colleagues in as far as it went beyond the common stock is more of a question. Two Florentine writers from the late 14th century will illustrate this.

One was Antonio de' Mazzinghi – author of the *Fioretti* that were spoken of above (p. 27). In the present context we may observe that he uses more or less the same symbols as Biagio though with extensions of the system – for instance[82]

$$\frac{r.1805.c.e.p \; 19\rho}{r.20.c} \quad \text{meaning} \quad \frac{radice \; 1805 \; censi \; e \; più \; 19\rho}{radice \; 20 \; censi} \quad \text{or} \quad \frac{\sqrt{(1805 \; censi)}{+}19\rho}{\sqrt{20} \; censi} \; .$$

Occasionally, like Biagio, Antonio still writes *censo* in full within the formal fractions; outside the formal fractions, *radice* is often abbreviated ℞. New is ℞ or an encircled fully written *radice* indicating that a root is to be taken of a binomial.[83] Dardi, instead, had used a verbal expression, ℞ *de zonto*, "root of, joined ...". Others, also after Antonio, would speak of a *radice generale*, *radice universale* or *radice legata* ("general", "universal" or "bound root"). Antonio combines the two, and uses writes for instance "general root of one *thing* plus root of 350 less one *censo*" (underlining represents encircling) where we would write

$$\sqrt{thing + \sqrt{350 - 1 \; censo}}$$

(more on this in the next chapter). Whereas the encircled *radice* is a notational but no conceptual innovation, the combination in nesting certainly was (and to my knowledge not emulated by others at the time).

The margin is sometimes used for summaries of what has been calculated in the text. It also serves, as we know it from Biagio and with traces in Gherardi, for addition of formal fractions with indication of the cross-multiplication – now, however, for instance on fol. 464[r]) with an extension. The text asks for the addition of $\frac{80-16\rho}{5+1\rho}$ and $\frac{80+16\rho}{5-1\rho}$ («–» here replaces Antonio's *mê*, «+» his *p*). In the margin we find

$$\frac{80-16\rho}{5+1\rho} \times \frac{80+16\rho}{5-1\rho}$$

and below this the determination of the resulting sum

```
400    80ρ
        80ρ     16c
400  +160ρ  +  16c
400  −160ρ  +  16c
800         +  32c
        25−c
```

The first three lines calculate $(5+1\rho)\cdot(80+16\rho)$ in a scheme borrowed from the multiplication *a scacchiera* ("on chessboard" – very close to our present algorithm) of

---

[82] This is precisely where Benedetto declares his intention to speak like Antonio (above, p. 27) after having pointed out that one might as well say $9\,[0]\frac{1}{2}+\sqrt{18}\,\frac{1}{20}$. There is thus no doubt that Benedetto follows Antonio precisely. We also observe that the notation used here is not *precisely* what we encountered in the borrowings from Biagio, confirming the intended fidelity.

[83] This encircling is left out in Arrighi's edition, it has to be found in the manuscript.

multi-digit numbers. The fourth line states the result of $(5-1\rho)\cdot(80-16\rho)$ without calculation, and the fifth the sum of the two products, that is, the numerator of the sum; the sixth, finally, contains its denominator (under a stroke that now serves as fraction line).

Nothing similar to this second part of the marginal calculation is found in the *Alcibra amuchabile*, in spite of its detailed explanation of the summation of the two formal fractions (nor in Benedetto's transcription of Biagio's problems, in spite of the three marginal indications of cross multiplications). We may therefore assume that the use of the *a scacchiera* scheme was a fresh development when Antonio wrote – perhaps but not necessarily due to Antonio himself.

The slightly later Florentine *Tratato sopra l'arte della arismetricha* (above, p. 28) is mainly of interest for the present chapter because of its use of schemes for polynomial arithmetic. These are written within indented frames, which however at times take up the whole column breadth.

Those which are simple enough (multiplications of two binomials) make use of the scheme which we encountered in *Alcibra amuchabile* and in more elaborate form in Dardi; here they are even less developed, the lines being omitted. Those that are more intricate are made *a scacchiera*, as we just encountered it in Antonio.

As current at the time, ℞ is used for *radice* in the running text as well as in the schemes. No other abbreviations are used in the running test, but within the schemes, the *cosa* may be either *chosa* or ρ, while the *censo* may be written in full or abbreviated *c*. Once more we see that rudimentary symbolic syntax does not require a symbolic lexicon.

A collective work from 1429

The di-Davizzo fragment (above, p. 13) contains no algebraic symbolism and no algebraic abbreviations beyond the use of ℞ for *radice*. The rest of the algebra within which it is contained (*Alchune ragione*) can be taken as a witness, if not of what was commonly done in the Florentine environment in 1424, at least of what was done by a group of writers probably connected to a single school or master.

℞ is used regularly. There are no formal fractions anywhere. Abbreviations for *censo/zenso* (*cen*, $\square$) and *chosa/cosa* (*co*; ρ is absent) are used but not systematically. Often they appear above the coefficient, as for instance $\overset{cen}{1}$, $\overset{co}{120}$ and $\overset{\square}{8}$; a number may (occasionally but rarely) be written $\overset{n}{100}{}^{o}$ or $\overset{n}{100}$. When abbreviated, *meno* ("less") is mostly *mê* but occasionally *m̂*. On fol. 38^{r} we thus find

$$10 \ m\hat{e} \ 1 \ cosa \ via \ 10 \ m\hat{e} \ \overset{co}{1} \ fa \ 100 \ e \ 1 \ censo \ m\hat{e} \ 20 \ chose$$
$$10 \ less \ 1 \ thing \ times \ 10 \ less \ \overset{thing}{1} \ makes \ 100 \ and \ 1 \ censo \ less \ 20 \ things$$

Occasionally, the margin contains summaries that are more consistent, but never formal calculations – for instance this equation written without equation sign (fol. 39^{v}):

$$\frac{10}{\underset{1}{co}} \qquad \frac{\overset{co}{5} \; e \; 1 p i \grave{u}}{1}$$

All in all, this compilation presents us with unsystematic use of a symbolic lexicon but no symbolic syntax.

The notation differs from what we have seen so far, but the principle to write the symbol above the coefficient reminds of the Maghreb notation. Given the absence of formal fractions or any other kind of symbolic calculation, however, a not unlikely inspiration must have been rather ineffectual.

The Florentine encyclopedias

As we have seen, the three Florentine encyclopedias (above, p. 21), though borrowing al-Khwārizmī's introductory algebra (in Guglielmo de Lunis's translation),[84] remain within their own branch of the abbacus tradition (the tradition encompassing Biagio and Antonio – above, p. 22) when it comes to algebraic notation, albeit with a minor change and a major expansion.

The minor change is the transformation of the *c* abbreviating *cosa* into a conventional glyph which we may render σ.[85]

The expansion is the introduction of a long stroke functioning as an equation sign. This long stroke is used for example in the Ottoboniano *Praticha* (fol. 331^v) in a problem which in hybrid symbols may be rendered

$$\frac{100}{1\rho}+\frac{100}{1\rho+7} = 40 \; .$$

In the margin, we first find the two formal fractions $\frac{100}{1\rho}$ and $\frac{100}{1\rho\,7}$, and then this scheme

$$
\begin{array}{ll}
100\rho & \\
\underline{100\rho} & \underline{700 \qquad\quad} \\
200\rho & 700 \\
1\sigma & \underline{7\rho \qquad\qquad} \\
& \qquad\qquad 40 \\[2mm]
200\rho \qquad 700 \; \text{——} \quad 40\sigma \; \langle 280\rho \rangle
\end{array}
$$

corresponding to the calculation

$$\frac{100}{1\rho}+\frac{100}{1\rho+7} = \frac{100\rho+100\cdot(\rho+7)}{(1\rho)\cdot(1\rho+7)} = \frac{100\rho+(100\rho+700)}{1\,censo+7\rho} = 40 \; .$$

The long stroke appears to be a recent invention. In a problem borrowed from Antonio's *Fioretti*, Benedetto's margin (fol. 456^r) indeed gives $\frac{1}{1\rho} \quad \frac{1}{1\rho\,1}$. Borrowing the same problem

---

[84] Namely, as an expression of Humanist piety – because it is older than what else is found, as expressed in Benedetto's *Praticha* (fol. 371^v).

[85] Benedetto, when introducing the abbreviations systematically (fol. 374^r) still uses *c* (elsewhere also sometimes *c*°, with no system), and adds something looking like ♭ for *cubo* and ♭r for *cubo relato*, a term for the fifth power perhaps introduced by Antonio (see above, p. 27).

but having no intention to "speak like Antonio", the Ottoboniano writer uses what may be assumed to be his own ways, and writes $\frac{1}{1\rho}$ —— $\frac{1}{1\rho\,1}$. That example, by the way, illustrates the wider function of the long stroke as a "confrontation sign" – the two fractions express the gain respectively loss of two trading partners.[86]

Within this particular tradition, as we see, there is some development of notations from Biagio until the encyclopedias. It is quite modest, however, and very slow – and nobody within this tradition nor in those to which *Tratato sopra l'arte dell arismeticha* and *Alchune ragione* belong, attempts to innovate systematically.[87]

The Modena manuscript

As in its explicit presentation of *gradi*, the Modena manuscript differs from what else we have seen by its abbreviations. .R. is sometimes used for *radice*, and for the powers the manuscript uses powers capital letters, surrounded by points and conspicuously larger than and distinct from the letters used in the text:[88]

.N. – .C. – .Z. alternating with .ς. – .Q. – .Zς. – ...

continued in agreement with the naming (above, p. 34). There are no symbolic calculations. It is possible that Peletier's use of a stylized ς for the second power is a borrowing (the shapes are very similar, with a greatly exaggerated cedilla), but it could also be a parallel invention coming from his term *çanse*. Apart from that, I have seen no later traces of the system used here.

Though systematically used and being different from mere letters, these glyphs remain abbreviations. The running text where they appear is easily expanded into a classical rhetorical algebra, similar to that of al-Khwārizmī. There are no operations directly at the level of symbols. In Nesselmann's terminology, the text is syncopated.

The shift from .Z. (which corresponds to the normal late 15th-century spelling) in the initial tables to .ς. meant as .Ç. (which, we remember, corresponds to the northern spelling in Dardi's 14th century) in the running text might suggest that the original from which this algebra is copied had used Ç, and that the copyist, at first intending to modernize in agreement with his own orthography, soon gave up this idea and followed his

---

[86] As we remember, the formation of a simplified equation was seen as a "confrontation"; for 15th-century writers, the two functions were hardly different,

[87] A marginal note on fol. 309[r] of the Ottoboniano *Praticha* uses a triangle for the cube and a double square for the *censo di censo*, and furthermore a double stroke as equation sign. It is in a different hand, however, and probably from the sixteenth century, and therefore does not change this conclusion.

[88] This distinction holds in the beginning, when the names and abbreviations are introduced. It tends to be forgotten afterwards – with one partial exception: .ς., rarely used in the introduction, becomes preponderant in the main text, not an ordinary ç.

model.[89] *If* so, this would mean that the system we see here was in fact developed at a time where the normal spelling was *çenso. If!*

Chuquet

Chuquet's notation for the powers is inseparable from his names, and were therefore already spoken above (p. 35). It looks modern but had no influence, and did not serve in symbolic calculation. There is no reason to speak more about it.

Giovanni del Sodo

Giovanni del Sodo's idiosyncratic naming of the powers was described above (p. 38). The glyphs he uses for them are certainly no less his own [Ghaligai 1521: 71^r] – see Figure 5. The best we can say about this system is probably that it provides further evidence for experimentation with systematic abbreviation.

We may presume that del Sodo used these glyphs also in his work with algebraic problems – so does in any case Ghaligai, also in occasional symbolic calculations.

Figure 5. From [Ghaligai 1521: 1']. Redrawn.

Algebra in Italian print

We know del Sodo's algebra from a printed book and not from his own writings. In other respects there is much more to say about Ghaligai's printed book, but not as regards his account of his master's algebraic notations.

Pacioli

Three of Pacioli's works seem to be relevant for the topic of the present chapter – one, however, only marginally.

As already said on p. 36, the chapter is lost from the manuscript *Suis carissimis disciplis* in which a systematic presentation of his notation could be expected. Instead, this notation has to read from the problems where he uses it, which at the same time shows us *how* he uses if.

---

[89] Van Egmond [1986] proposes the inverse process. My general experience with copyists (or modern copy editors) who try to change a text is that they set out being more or less systematic and then give up. Moreover, part of Van Egmond's argument builds on a supposed coincidence of the writings of "3" and "z". He evidently had no magnifying glass at his disposal when working on the manuscript. See the depiction in [Høyrup 2010:40] – scalable microfilms or scans are sometimes better tools than the eyes applied to the original.

The *cosa* is indicated by a superscript ^{co}, while the *censo* is a superscript $\square$ and the *cubo* a superscript $\triangle$ – all written after the coefficient; higher powers are sparsely used, but scattered problems and explanations show that *censo di censi* is abbreviated $\square\!\square$, while *censo di cubi* becomes $\square\triangle$; neither *primo relato* nor *secondo relato* are abbreviated. The notation is used in simple marginal calculations (fols 117^v and 118^v, etc.); fol. 250^r, a simple symbolic equation (written with the long stroke as equation sign) goes through stepwise transformation in the margin; on fol. 251^v simple formal fractions are used within the running text. Mostly, however, the exposition is syncopated.

There can be no doubts that this notation descends from the one we know from the *Alchune ragione* (above, p. 53)

The *Summa* was printed with extremely small distance between the lines. It would have been next to impossible to use superscript glyphs or abbreviations, which is probably the reason that Pacioli now changes his notation, as indicated on p. 37, making use of abbreviations and not of non-letter glyphs. The typesetting also does not allow for the writing of formal fractions within the text column. They might still have been placed in the ample margin, but the margin is only used for diagrams (and only in the geometric part). Equation manipulations *could* also have been shown in the margin, but they are not.

Pacioli knew about formal fractions as well as symbolic equations and their transformation – *Suis carissimis* leaves no doubt about that. He evidently did not find these techniques so important that he would ask the printer to show them in his *magnus opus*.

A third way to denote algebraic powers is chosen by Pacioli in his translation of Piero della Francesca's *Libellus de quinque corporibus regularibus*, printed as part III of *Divina proportione* in [1509] by the same printer as the *Summa*. A comparison with Piero's own manuscript version (Vatican, Urb. lat. 632) demonstrates that Pacioli really makes a deliberate choice. Piero, writing in Latin, speaks of the first power as *res*, and overscores the coefficient. "9 *things*" thus appears on fol. 4^v as "$\bar{9}$ *res*" (but *res* may be left out). His second power is *census*, represented by a square $\square$ written over the coefficient[90] – close to what Pacioli had done in *Suis carissimis*.

In the corresponding passage (fol. I^v, and elsewhere), Pacioli uses $\diamondsuit$ written on the same line and after the coefficient for the *cosa* and $\square$ for the *censo*. These glyphs are not used in symbolic calculations; Pacioli's preference for the use of mere abbreviations is revealed on fol. 3^v of part I of the treatise, where he explains that various professions, among whom *le mathematici per algebra*, use specific *caratheri e abreviature* "in order to avoid prolixity in writing and also of reading".[91]

---

[90] In his algebra compilation (BNC, Conventi Soppressi A.VI.2606) Piero does as much (with *cosa* instead of *res*) – e.g., fol. 46^v; even less systematically, however.

[91] A table in the manuscript (Milan, Biblioteca Ambrosiana, Ms. 170 Sup., written in 1498) specifies. It lists abbreviations and glyphs for *radice, più, meno, quadrato* (*cosa* and *censo* are absent), together

Three 16th-century Italian writers

Tartaglia [1560: 1ʳ] repeats not only Pacioli's names and numbering for the *dignità* (above, p. 39) but also his abbreviations. He uses them, however, in symbolic calculations when dealing with the arithmetic of binomials, and also in formal fractions; but he does not go beyond what was already done in Florence in the early 15th century.

Cardano is less systematic in the *Practica arithmetice et mensurandi singularis* from [1539], but from time to time he uses the same abbreviations as Pacioli (even *co* for the first power, although his full Latin term is *res*). He also uses the abbreviations in occasional symbolic calculations. The *Ars magna* from [1545] is similar. Cardano was a great algebraist not *because of* his notation and occasional symbolic calculations but *in spite* of what he did on this account.

Bombelli's notation for the powers is more innovative than his names. Along with his list of names [Bombelli 1572: 204] he shows the abbreviation he is going to use, namely the exponent written above an arc – the 10th power, for instance, being 1̆0̆.

This allows him to formulate rules for multiplication and division of powers. When negative powers result, as in the division of 20 by 4̬, Bombelli proposes "20 *esimo di* 4̬", corresponding to "20 -th of 4̬" (just as ungrammatical in Italian as in English), but alternatively by means of a formal fraction.

Bombelli's system is similar to that of Chuquet, but with the difference that Chuquet used the exponent also for roots. Bombelli, instead, uses letter abbreviations (p. 6):

R.q. (*radice quadrata*) – R.c. (*radice cubica*) – RR.q – R.p.r – ...

Bombelli's notation had some influence; Simon Stevin [1585a: 26] knew him as "a great arithmetician of our time", and apparently borrowed his notation, although replacing the arc by a full circle around the exponent. In *De thiende* [1585b] he used the same encircled numbers to indicate places in decimal fractions; a borrowing in one or the other direction is certain. The change to Bombelli's notation may have been made because the printer (Christofle Plantin in both cases) already had the encircled numbers at hand.

But even that influence resulted just as much a dead end as Chuquet's forgotten manuscript. The ink in Stevin's *Arithmétique* was hardly dry when Viète undertook to give algebra a new shape which would soon make Stevin's approach irrelevant.

German notations

The transfer of algebra from manuscript to print thus did not bring any fundamental change to the use of symbolism in Italy, in neither quality nor density of use. To the contrary, the transfer to German lands soon did.

(*inter alia*) with abbreviations for *linea, geometria* and *arithmetica*) – see [Maia Bertato 2008: 13].

The German and Latin algebras of Dresden, C 80

No immediately, however. During the initial eclectic phase, the use of notations was equally eclectic.

We may leave aside the notations used by Regiomontanus in his private notes and manuscripts. They are all but uniform (see, for example, [Høyrup 2019a: 338]), and reflect the variety of sources he was using.

More interesting are the "German" and the "Latin algebra" contained in the manuscript Dresden, C 80 (above, p. 39). An edition of the former was made by Kurt Vogel in [1981]. As pointed out by Vogel (p. 11), the number "unit" is indicated in six different ways, one of which is an abbreviation and two are non-letter glyphs. The first power can also be written in six different ways, two of which can be regarded as abbreviations while one is a non-letter glyph. Even the second to fourth powers are represented in several though fewer ways. Some of these names are root names, others not. At times, the pseudo-fraction notation we know from Dardi (above, p. 47) is used (occasionally inverting the roles of "denominator" and "numerator"); but in the very end a formal fraction appears (no Italian source I know of had ever mixed these two). Apart from this single formal fraction, no symbolic calculations are made.

The Latin algebra which is contained in the same manuscript may have served Widmann for his algebra lectures (above, p. 39). As said, the treatise is a conglomerate, but its constituents overlap and supplement each other. As also said, it uses a set of non-letter glyphs (mostly more or less standardized stylizations of letter abbreviations) so consistently that it is difficult to find out what the intended full verbal names are:

ϕ	*numerus*
$\mathcal{X}$	*res*
$\mathcal{z}$	*zensum*
$\mathcal{P}$	*cubus*
$\mathcal{z}\mathcal{z}$	*zensum zensorum*

Less systematic is the use of a point «·» for *radix*, "root", and «··» for "root of root". Cube root, instead, when not written in full, is "℞ *cubica*"; the square root also occasionally appears as ℞.

For those who are interested in the etymology of mathematical symbols it is worth noticing that a mere point when written on uneven paper with a pen may be difficult to distinguish (and to make). It was soon written with a down- and an upstroke, as ̌, which gave rise to the modern root sign √.

While the Italians, when abbreviating *più* ("more", whence "plus") and *meno* ("less", whence "minus"), had used letter abbreviations, the present manuscript uses + and –. Like the transformed dot, even these glyphs, mostly serving as symbols, were also taken over by others and ultimately by modern mathematics.

Formal fractions are regularly used where they can serve. Beyond that, there are neither occasions for nor use of symbolic calculations; on the whole the presentation is syncopated.

Alexander

Alexander, as we have seen (above, p. 41) has names for powers until the ninth with corresponding glyphs, and he uses the latter systematically. He conserves what is found in the Latin algebra just discussed, and expands it into what was to become the canonical sequence of the *coβ*:[92]

φ	1.	*dragma* or *numerus*
ҳ	2.	*res*
ҙ	3.	*zensus*
cʳ	4.	*cubus*
ҙҙ	5.	*census de censo*
ß	6.	*sursolidum*
ҙcʳ	7.	*censicubus*
bß	8.	*bissursolidum*
ҙҙҙ	9.	*census censu de censu*
cɼʳ	10.	*cubus de cubo*

As we see, Alexander falls into the same trap as Pacioli, numbering the powers in a way that does not coincide with our exponents, encumbering the reduction of equations involving higher powers (which does not prevent him from performing these reductions correctly[93]).

The arithmetic of binomials (addition, subtraction, multiplication) is shown in schemes emulating those for operations in Hindu-Arabic arithmetic. The operations on formal fractions are also explained.

The *Initium algebrae* discards the inadequate numbering of the powers.[94] As regards its use of schemes and formal fractions it is similar to Alexander's Latin manuscript, but it goes a bit further (without persevering in these experiments), showing [ed. Curtze 1902: 517] how to multiply two formal fractions, whose numerators themselves are formal fractions.

---

[92] As far as I can read in the manuscript, Alexander cannot agree with himself whether the ablative of *census* (4th declination) should be *censu* or *censo*. His school grammar may have offered no better assistance than mine, which also gives *frūctū* but *domō*. He probably knew his mathematics better than the petty details of his grammar.

[93] Accordingly, on fol. 46ʳ, a tabulation of the products of powers does not include this numeration but only the corresponding powers of 2. In this tabulation, contracted writings for powers until the 18th are used – when possible, by glyphs combined by embedding, when not by numbered *sursolida* – the 17th power is thus *quintß*.

[94] From this we may probably conclude that it is based on a later version of Alexander's Latin manuscript than the one we possess.

Schreyber, Rudolff and later *coß*

Heinrich Schreyber's naming of the powers by their exponents was spoken of above (p. 41). He also knows about the use of schemes in polynomial arithmetic, and about operations with formal fractions. He may have learned from Alexander. In the present context there is nothing more to say about his book.

Rudolff certainly learned from Alexander (above, p. 41), and since his book was the one that came to define *coß* everywhere and for as long as algebra carried this name, he is the principal source from whom the later *coß* tradition until Mennher and Clavius learned the symbols for the first nine powers, with minor variations only depending on how printers chose to cut them; he also taught later cossists their use in schematic polynomial arithmetic and formal fractions. As regards the general adoption of +, − and √, they were probably already so widely spread (except perhaps √) that Rudolff was not needed.

Rudolff's scheme for the products of powers (fol. 28^v) is triangular, not quadratic, and therefore does not produce powers above the cube of cube; on the other hand he changes the numbering, letting *number* correspond to 0, and thus gets exponents.

German algebra thus did not *discover* the use of algebraic abbreviations and other glyphs, nor their use in symbolic calculation. All the German writers do had already been done in abbacus algebra. But from the Latin algebra in C 80 onward, German algebra discovered that this kind of notation is most useful if used consistently. If we hark back to Woepcke's discovery of the Maghreb notation [1854: 355], he declares that[95]

> the indispensable condition for characterizing a set of conventional signs as a notation is that they are always used when it is fitting, and always in the same way.

The "notations" of the abbacus writers would thus not have been accepted by Woepcke as algebraic notations, nor those of Regiomontanus and his German contemporaries. The phase shift only occurred a small generation later, with the writing of the Latin algebra in C 80.

French algebraic writings

The French scene is very different − it still recalls Pacioli's "as many heads, so many opinions". Let us first look at de la Roche (cf. above, p. 35).

When introducing the algebraic powers [1520: 42^r], he refers to a variety of names for them, using examples with coefficients 12 and 13.

Nombres linéaires: 12.^{1} or 12.ρ
Nombres superficiels quarrés: 12.^{2} or 12.$\mathcal{P}$

---

[95] "la condition indispensable pour donner à des signes conventionnels quelconques le caractère d'une notation, c'est qu'ils soient toujours employés quand il y a lieu, et toujours de la même manière".

Nombres cubiques: 12.³ or 12.□

Nombres quarrés de quarrés: 12.⁴ or 12.ᑕᖶᑭ

In what follows, de la Roche is going to use the notation to the right: ρ, ᑕᑭ, □, ᑕᖶᑭ. ρ is almost certainly borrowed from the Florentine tradition, but ᑕᑭ and ᑕᖶᑭ point to early German algebra. ᑕᎰ is indeed used for the second power in a Latin section (fol. 40ʳ⁻ᵛ) of the manuscript Leipzig, Clm 1409, which is dated 1459. It is an abbreviation for *ce* (it serves elsewhere for *centenarius*, which leaves no doubt).[96] For "plus" he uses ṗ, for "minus" ṁ. *racine* is written ℞. These three could have been taken over from Chuquet.

Apart from that, it is not at all clear exactly how de la Roche got this notation (but Lyon was a city with much international commerce). What seems clear is that his notations went nowhere. Peletier (above, p. 42), who carefully listed not only the books he knew but also those he knew *about* without having seen them, does not mention him in [1554] (in the Latin version [1560: *ivᵛ] he does, however). As to Peletier's notations, he mostly follows Stifel (and thus Rudolff). The first power, however, he calls *racine* and denotes ℞, and he keeps the abbreviations *p* and *m* for addition and subtraction – introducing the term *exposant*, "exponent" (as already told on p. 42). As Stifel, he uses the notation for symbolic calculations in schemes and formal fractions, without going beyond him.

Jean Borrel [Buteo 1559: 123], dealing with three powers only, uses ρ for the first power (he knows de la Roche and is more likely to have borrowed from him than from some Tuscan manuscript). The second power he abbreviates ◇, and the third ⬠ – both probably his own inventions. Addition and subtraction are indicated by *P* and *M*. There are a few schemes showing the arithmetic of polynomials. All in all, nothing beyond what could be expected in a primer, and well below what Peletier had found in Stifel (or what he will have found himself in de la Roche).

Gosselin (above, p. 43), as mentioned, borrows some but not all names and abbreviations for the powers from Xylander's translation of Diophantos. Diophantos, followed by Xylander, constructs higher powers multiplicatively, while Gosselin produces them by embedding (left from [Xylander 1575: 1], right from [Gosselin 1577: 4ᵛ–5ʳ], who comments on the differences):

numerus, N	*latus*, L
quadratus, Q	*quadratus*, Q
cubus, C	*cubus*, C
quadratoquadratum, QQ	*quadratoquadratum*, QQ
quadratocubus, QC	*relatum primum*, RP
cubocubus, CC	*quadratocubus*, QC
	relatum secundum, RS
	(omitted, elsewhere) QQQ
	cubocubus, CC

---

[96] In the same section, the first power is written ʒ₂, a variant of ꝛ.

Xylander's *numerus*, translating ἀριθμός, is evidently in conflict with the inherited use of *number* in the algebraic tradition. Since Xylander explains it to be *latus quadrati*, Gosselin may have taken *latus* from him.[97]

Gosselin uses this notation consistently, also in schematic polynomial arithmetic and formal fractions, but not for other kind of symbolic calculation. The continuation until exactly the ninth power suggests an unconfessed inspiration from cossic writings.

Viète's naming of powers [1591a: 4ᵛ] ("powers" in general, not the powers of *the* unknown since he has several) is taken over from Xylander's Diophantos:

The first is	*latus*, or *radix*
2	*quadratum*
3	*cubus*
4	*quadrato-quadratum*
5	*quadrato-cubus*
6	*cubo-cubus*
7	*quadrato-quadrato-cubus*

The name for the seventh power goes beyond Xylander, but follows automatically from the principle that produced the preceding ones. *latus* could come from Xylander (cf. above, p. 63), but could also be a borrowing from Gosselin, or even from Cardano's *Ars magna*. *radix* could come from many writings from the earlier tradition – since Viète only cites predecessors in order to blame we cannot know from where.

Viète still does not speak about "powers", but on fol. 5ʳ he introduces the term that would eventually take on this meaning. *Potestas* designates instead the dimension of any product of powers of different magnitudes; if (in our terms) it is a power of a single magnitude, it is called "pure".[98]

"Eventually", however, was not far away. The term surfaces again in *Ad logisticem speciosam notae priores* (originally planned to appear first after the *Isagoge* but only published in 1631 [Witmer 1983: 9][99]). Here [ed. van Schooten 1546: 14], *potestas* has taken on the modern meaning.

---

[97] There is thus no need to postulate a borrowing from Ramus's (anonymous and rather trite) *Algebra* [1560: 2ʳ], whose notation is otherwise quite different.

[98] The notion of algebraic "powers" thus does not descend by generalization from the Diophantos's use of δύναμις for the second power of the unknown and the related Euclidean usage in *Elements* X, in spite of its similar translation *potentia*. Indirect inspiration is also improbable – Viète would have known Xylander's translation *facultas* [1575: 1]. The later French choice of *puissance*, however, may well have been well have been influenced by the Diophantine usage. English "power" might come from the French term; in [1570: 60], however, Henry Billingsley uses "power" for the geometric δύναμις. Thomas Harriot [1631: 1], writing in Latin, takes over Viète's terms, *potestas pura vel affecta*.

[99] Since Viète withheld the text, we can evidently not be sure that what was published in 1631 was already formulated in 1591; any date before Viète's death in 1603 is possible.

Each single of these French writers uses his notation in a way which would make Woepcke accept it as a genuine algebraic notation. But each of them follows his own way – it is not possible to put them together as a "French 17th-century algebraic tradition" or "school" comparable to the *coß*.[100]

---

[100] Basing himself on different criteria François Loget [2012] reaches the same conclusion.

Chapter IV. Embedding and parenthesis function

The importance of the parenthesis was pointed out in the Introduction. A parenthesis is what allows a whole algebraic expression to serve as the argument for a function (speaking in modern terms) – to be *embedded*.

A simple variety of embedding was spoken of repeatedly in Chapter III – namely when, for instance, the sixth power of the unknown was spoken of as the "*censo* of *cubo*" – in modern terms, once again, as $(t^3)^2$. This way of speaking, and of understanding things, started without being systematic with Antonio de' Mazzinghi, and took over almost completely in the late 15th century, a hundred years after Antonio (Viète's use of the multiplicative principle in 1591 being an absolute exception, imposed by his determination to follow Diophantos).

From Pacioli onward we may therefore be tempted to see the powers as "functions"; but if we do so we shall be aware that these "functions" allow only a very particular kind of arguments – namely, other powers. We never see the *censo* of a binomial, a formal fraction, or a root.

This kind of embedding, however, is only one kind of parenthesis. Other types can also be found – we may be tempted to consider them "primitive" (in the sense of "early") but should rather see them as special-purpose parentheses.

The first special-purpose parenthesis which we find in abbacus algebra is found in the formal fractions. This was already used in the Maghreb notations (above, p. 44), where the transformed abbreviation for division could serve as a fraction line. In a formal fraction, indeed, the numerator as well as the denominator *are* parentheses. As we have seen in Chapter III, the use of such formal fractions is already reflected indirectly in Gherardi's *Libro di ragioni* from 1327, and they are used regularly throughout the later tradition.

Composite radicands

Another special-purpose parenthesis serves to take the root of a composite radicand. It is expressed in varying ways.

Dardi

The first to make use of it seems to be Dardi. On fol. 9^v of the Vatican manuscript we find, for instance, "℞ *de zonto* $\frac{1}{4}$ *cô* ℞ *de* 12" (standing for $\sqrt{\frac{1}{4} + \sqrt{12}}$); *zonto* corresponds to Tuscan *gionto*, "joined", and the whole expression thus means "root of, joined $\frac{1}{4}$ with root of 12"). Others, later on, would speak of a *radice generale*, a *radice universale* or a *radice legata* ("general", "universal" or "bound root"), mostly with the same meaning.

Antonio – two levels

As long as we are sure that the radicand is a simple binomial, this is unambiguous. Antonio, as we have seen (above, p. 52), uses ⑬ or an encircled *radice* in the same function – but since his radicands are sometimes more complex, he uses "general root" to indicate the "outer brackets", writing for instance "general root of one *thing* plus <u>root</u> of 350 less one *censo*" (underlining represents encircling) corresponding to our

$$\sqrt{thing + \sqrt{350 - 1 \ censo}} \ .$$

⑬ was taken over by Antonio's student Giovanni di Bartolo. In a section of the Palatino *Praticha* which is said to contain problems of his (fol. 470^r–478^v), it is first explained (fol. 473^r) that it means that the root "has to be taken of everything that follows". In that passage the example "$\underline{\mathbf{R}}$ 10 less 2 *things*" is given. Later on (fol. 477^v) it is used repeatedly to take the root of "25 less 10 *things* less 3 *censi*", that is, of a trinomial.[101] This is different from the root of a binomial which we know from Dardi, but not part of a two-level system. In order to be sure what is means, one has to follow the calculations with full attention.

Chuquet and de la Roche

On fol. 52^r of his *Triparty*, Chuquet introduces the notion of a *racine lyée*, "bound root", corresponding to *radice legata*. He explains it by the example $\mathbf{R}.^2\underline{6.\bar{p}.\mathbf{R}.^27.}$ *lyée d'une ligne par dessoubz*, "bound by a line below". His explanation is less than satisfactory, "there where before 7 was a second root, now it is $\mathbf{R}^2$ of $\mathbf{R}^2$, that is, fourth root; and 6 is $\mathbf{R}$ which before was number". However, in the copious calculations that follow he does not fall into his own trap.[102] Beyond the "bound roots" he speaks (fol. 63^r) of "composite roots" (*racines composées*), for instance $\mathbf{R}.^2\underline{15.\bar{p}.\mathbf{R}^2.10.}\bar{p}\mathbf{R}^2\underline{13.\bar{p}.\mathbf{R}^26.}$ The verbal distinction does not correspond to a conceptual difference, the same symbolism covers both unambiguously.

Chuquet's notation would have been adequate for repeated nesting: nothing would have prevented a double underlining in the manuscript. If I am not mistaken, Chuquet never takes advantage of this possibility – probably because he never had the need, just as simpler abbacus algebra rarely needed to specify the range of its joined roots because

---

[101] An edition of this problem collection is contained in [Arrighi 1967b]. Unfortunately, Arrighi does not distinguish $\mathbf{R}$ from ⑬ in his transcription, for which reason the manuscript has to be relied upon.

[102] Others were less fortunate. The *Libro di conti e mercatanzie* [ed. [Gregori & Grugnetti 1998: 116] thus presumes in earnest – in perfect agreement with Chuquet's explanation – that $\sqrt{(a+\sqrt{b})} = 6\sqrt{a} + \sqrt{\sqrt{b}}$.

it rarely encountered radicands that were not either monomials or binomials. Why "take advantage" if no advantage is offered?

He also does not take advantage of the possibility to use the notation for taking composite expressions as arguments for other "functions". That had an obvious reason: As long as algebraic powers were not treated as functions that could take any algebraic expression as their argument, there *were* no such other functions. (When a product of two polynomials was needed, it could be calculated separately, rhetorically or in a scheme.)

De la Roche takes over Chuquet's term (fol. 30^v) but not his notation – that is, the reader has to understand from context when a bound root is meant. De la Roche himself seems not to have understood too well. As an example, he shows how to make $\sqrt{(5+\sqrt{3})}$ and $^3\sqrt{(4+^3\sqrt{7})}$ comparable, but in the calculations replaces the latter with $^3\sqrt{(4+\sqrt{7})}$. Then, in order to make the two roots comparable he transforms both into sixth roots by calculating $(5+\sqrt{3})^3$ and $(4+\sqrt{7})^2$ – unfortunately finding the first to be $170+\sqrt{7500}+\sqrt{2352}$ instead of $170+\sqrt{21168}$, and the second to be $22+\sqrt{384}$ instead of $22+\sqrt{448}$.

He has the good judgement not to persevere, giving the reason that the topic "serves little in our answers to the questions".

Mennher

Valentin Mennher's *Arithmetique seconde* was printed in Antwerpen in [1556]. Mennher was born in Kempten in Schwaben in 1520.[103] When young he worked as an accountant in the Fugger firm, and in Fugger service he came to Antwerpen, where he later opened a school. In 1550 he had promised this "second arithmetic" (thus stated in the preface in 1556 – the promise must have been made in a first *Arithmetic* that is now lost).

The *Arithmetique seconde* consists of three parts. The first is a regular and well-structured *Rechenbuch* in German style, the second an algebra, the third a geometry going well beyond traditional abbacus geometries, containing both a Euclidean *proof* of the Pythagorean theorem (fol. S i^r) and an Archimedean determination of the ratio between the perimeter and the diameter of a circle.[104]

The second part, about algebra, explains on fol. F iiii^r that understanding of "the high and liberal art of arithmetic is infinite"; several questions, moreover, "cannot be solved except by the very ingenious rule of algebra, or *cos*; as also commanded by the very subtle and liberal art of geometry". The "style and manner" of the very renowned Christoph

---

[103] See [Meskens 2013: 14*f* and *passim*], where Mennher is discussed copiously.

[104] Traditional abbacus- and *Rechenmeister* geometries had always accepted the approximation 22 : 7 as a quasi-axiom (as does Mennher himself in the preceding pages).

It is worth noticing that in [1564] Mennher published the "practice of spherical triangles, the distances on the globe, clocks, shadows, and other ingenious and new mathematical questions" – adding, thus Mennher's preamble, to what had been done by the very learned Regiomontanus (*viz*, in *De triangulis*) the labour of calculation.

Rudolff has been of great help, and Mennher has found him very competent. Therefore, he says, he has not deviated much from him, knowing well that the very renowned Michael Stifel has renewed and augmented him much in the same High German language with several beautiful examples – from which, however, "I have extracted a fair part of the best only, adding other matters needed by merchants".

Mennher [1556: Fv^v] refers to universal as well as bound roots (*universel*, V., respectively *lié*, L.). One might suspect a terminological borrowing from de la Roche, but the suspicion seems not to be justified. On fol. Hvi^{r-v}, √4+√16 and √3+√2 serve to exemplify the *racine liée;* the former is determined as 2+4, showing that the *racine liée* is nothing but the sum of separate roots, while the second cannot be reduced. The universal root is exemplified by V.√20+√25, which is calculated as √(20+√25) = √(20+5) = √25 = 5; it is thus identical with Dardi's "joined root".

This meaning of the *racine universelle* is confirmed on fol. Pi^r. It is used if not often then repeatedly in the book, as also in [Mennher 1565]. The *racine lié* is less appealed to in both works – they just write the sum of roots without using the superfluous concept, which Mennher may have felt obliged to introduce after having picked it up from somewhere (if from de la Roche, understanding badly what was meant; but the misunderstanding may have been the responsibility of an intermediary – where we can control him, Mennher is an intelligent reader). However that may be, Mennher understood what he was doing himself.

Cardano, Tartaglia and Bombelli

In the *Ars magna*, Cardano needed the same nested structure as once Antonio. His *radix universalis*, abbreviated ℞V is the usual "bound root" – on fol. 9^v, his ℞V:*cubica*℞19:p:3 is thus ^{3}√(√19+3). On fol. 14^r, where the nested structure is needed, he introduces a new term: "℞*universalissima* 10 p:℞V:100 m:4 *pos*" stands for √(10+√(100–4t)).[105] *radix universalissima* thus corresponds to Antonio's "general root", and Cardano's ℞V to Antonio's ⊛ (cf. above, p. 52).

The cumbersome term leaves little doubt that Cardano had had to reinvent. In the second edition [Cardano 1570: 26], he replaces *universalissima* with *tota* – inserting, however, a parenthesis "*(quam universalissimam appellare solent)*", "(which was habitually called most universal)".[106]

---

[105] *pos* stands for the "position", that is, for the first power of the unknown.

[106] Cardano, as we see, knew well what we may call the "rhetorical parenthesis", an aside in a written text. That was not new, this parenthesis in this shape was in use at least since the late 15th century (Pacioli employs it in the *Summa* in 1494). [Parkes 2016: 314] shows another example from the same year. Different shapes can be found in manuscripts from the late 14th century (*ibid.*, p. 213, 305).

Florian Cajori states [1928: I, 392] that Cardano uses round brackets a single time in the *Ars*

Tartaglia may have been the first to employ round brackets to delimit algebraic parentheses – perhaps for disambiguation of universal roots but in any case without system. He introduces these roots in *La seconda parte del general trattato* [1556: 81^v], explaining the meaning of "RV.8 men R60" (without the brackets). Returning to the topic on fol. 168^r he introduces a universal root of the trinomial $10+\sqrt{7}+\sqrt{5}$, writing it "RV.10 piu R7. piu R5" (still without brackets). For a while he works with this ambiguous notation, but on fol. 167^v he states the sum of 12 and "RV.20 *piu* R6" to be "12 *piu* RV.(20 *piu* R6")". This is used a few times, and then on fol. 169^r Tartaglia returns to his old ways, using the left round bracket (with no right counterpart) for other purposes. If disambiguation was aimed at, the attempt was not successful. Clavius [1608: 132–134, 159] uses the round brackets in such expressions as "$\sqrt[3]{}(12+\sqrt[3]{}32)+\sqrt[3]{}(12-\sqrt[3]{}32)$" and is likely to have picked them up directly or indirectly from Tartaglia.

Bombelli's *L'algebra* [1572] was presented in general as an attempt to improve Cardano's exposition in the *Ars magna*, "in which he explained much about this science [algebra], but obscurely in the saying" (5th page of the unpaginated letter "to the readers"). He tries (successfully) to do so on the present account, too.

Bombelli prefers the name *radice legata* but also knows *radice universale* (p. 99). If the radicand is a binomial, he mostly abbreviates R.q *legata* or R.c *legata* for bound square respectively cube root. If the radicand is a tri- or higher polynomial (and sometimes if it is a binomial), he encloses it between crotchets, an initial L and a final inverted ⅃ (dropping the word *legata*). The system allows nesting, for instance (p. 355) with three levels

R.q.L R.c.L 4608.p.R.q. 4456448⅃ .p.R.c.L4608.m.R.q.4456448⅃ .p.16⅃

corresponding to

$$\sqrt{\sqrt[3]{4608+\sqrt{4456448}}\ +\ \sqrt[3]{4608-\sqrt{4456558}}\ +16}$$

We might believe the initial crotchet L to be an abbreviation of *legata*, and that may perhaps have been the typesetters idea. It is not what Bombelli thought – in the manuscript Bombelli had used "beds" ⌞   ⌟ which made the nesting visually much more obvious than in the print – see the facsimiles in [Bortolotti 1929: 6, 9]. That cannot have been easy for the printer, who therefore showed only the corners of the beds.

---

magna. His reference [Cardano 1663: 438] turns out to be instead to the *Sermo de plus & minus*, a manuscript which was never published by Cardano himself but written in response to Bombelli's *L'algebra*, that is, no earlier than 1572 [Confalonieri 2013: 338]. It is not at all clear that an algebraic parenthesis should be meant – the brackets are used in a scheme, and enclose (with a misprint, 10 instead of 1) an expression that is to be squared, and below (after a dubious calculation) the result, as also stated in the text. The whole thing has to do with Cardano's vain struggle with Bombelli's imaginary numbers.

So, when he needed it, Bombelli could invent notations;[107] but he did not get the idea to use this parenthesis for purposes beyond the representation of complicated radicands – and probably had no occasion to get it because he had no use for it.

The substitutes for and a dubious step toward a general parenthesis function

It happened, of course, that abbacus masters formulated problems or performed calculations where we would find it natural to make use of parentheses. Then they did exactly as we do when making calculations on a pocket calculator or when writing a simple computer problem: they calculated what we would make a parenthesis, and saved the outcome, retrieving it later when it had to be used in further calculations. We may look at one example (there are many), taken from Pacioli's *Summa* [1494: 107^v]:

> If a quantity be divided into 2 parts, which are mutually divided; and the two results are joined together; and save the sum. And then, if you square each of the said parts; and the squares joined together; and this sum divided in the saved sum; from which shall come a determined number. I say that who makes of the first quantity two parts; where the surface of one in the other makes the said number; will always have the said parts.

In symbols indeed

$$\frac{a^2+b^2}{\frac{a}{b}+\frac{b}{a}} = \frac{ab\cdot(a^2+b^2)}{a^2+b^2} = ab \ .$$

As we see, the sum $^a/_b + {^b/_a}$ is kept together as one number by being saved and then, when it is to be used, retrieved.[108]

When polynomials were multiplied in schemes *a schacchiera*, each polynomial was written in a single line; for instance, in the Florentine *Tratato* [ed. Franci & Pancanti 1988: 11]

> 6 *things* and 8 and ℞9
> 6 *things* and 8 and ℞9

Here, each line can also be considered a special-purpose parenthesis.

---

[107] We may also think of his new notation for positive and negative imaginary numbers, "*più di meno*" and "*meno di meno*" – for instance [Bombelli 1572: 170], "2.p.di m. 2" where Euler would write $2+2\sqrt{-1}$ and we perhaps $2+2i$.

[108] We may remember the tentative iterated formal fractions from *Initium algebra* (not much later than Pacioli). If this principle had been unfolded, it *could* have been used to formulate Pacioli's theorem; but it would have been cumbersome, and the reduction of the denominator would not have been easier. There was certainly no incentive to develop whatever hunch Pacioli may have had into a workable technique.

All in all, abbacus algebra developed a small handful of special-purpose parentheses before 1500 (and then left matters there):
- the naming of powers by embedding;
- the use of formal fractions;
- ways to keep together composite radicands;
- the substitute-parenthesis achieved by saving and retrieving;
- the use of schemes to handle complicated calculations

but it never approached the unification of these into a single technique or concept.

When needed, the abbacus writers could develop new tools; as we have seen, Antonio created a way to express a nested parenthesis, and in the next chapter we shall encounter other examples of creativity. We may suppose that the need to create a general parenthesis did not materialize within the mathematical practice in which they were involved.

Viète

Viète's 17th-century editor and translators appear to have believed that Viète had got the idea of a the general parenthesis; at least they presupposed it in what they produced. The original editions of Viète's work shows that this is a mistake, a 17th-century reinterpretation of late-16th-century algebra.

In *Zeteticorum libri V* [Viète 1591b], a notation is occasionally used which with hindsight could be understood as an algebraic parenthesis. Curly brackets, or a single brace, serve to indicate that an expression going over several lines is meant to be kept together – but not in order to be operated upon as a whole. So, on p. 3^r we find the upper part of Figure 6; van Schooten [1646: 45] sees that there is no need for specification of a parenthesis – the fraction line suffices – and writes $\frac{B\ in\ H,\ =\ B\ in\ A}{F}$. Vasset [1630: 50] and Vaulezard [1630: 38] offer something very similar in their translations.

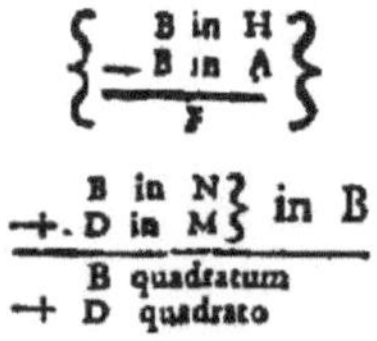

Figure 6. Viète's quasi-parentheses.

On Viète's fol. 18^r we see that a sole left brace can be used in the same function, and once again van Schooten (p. 70), Vasset (p. 142) and Vaulezard (p. 167) simply write numerator and denominator on a single line each. On Viète's fol. 17^r we see that a single right brace may also stand along the numerator alone – and even here, van Schooten (p. 69), Vasset (p. 138) and Vaulezard (p. 161) simply write the numerator in one line.

On Viète's fol. 15^r, on the other hand, we find something which the 17th-century editors and translators would see differently, even though nothing in Viète's original text suggests *he* saw any difference – namely the lower part of Figure 6. In this case, van Schooten (p. 65) and Vaulezard (p. 139) conserve the brace, seeing that it has a function. Vasset (p. 125) conserves the bracket but locates it after the fraction (containing both numerator and denominator), where it is actually superfluous; but he must have understood it as having a function in Viète's text (the mistake may have been the responsibility of the printer).

It is tempting to see this reinterpretation of Viète's text as a reflection of a pull produced by the development of 17th-century mathematics (it is noteworthy that both Vasset and Vaulezard wrote before Descartes): once the need for a more general parenthesis function was there, it made Vasset and Vaulezard read it into the text under their eyes, eliminating those braces that were superfluous and conserving those that had a parenthesis-defining function. Van Schooten's situation is evidently different, which makes him less significant: he was close to Descartes; he wrote after 1637; and he had been involved in the preparation of Descartes' *Geometrie* well before it was published [van Randenborgh 2012].

In any case, Vasset's and Vaulezard's readings of Viète reflect a need but could not fulfil it. That had to wait, not only until Descartes but, as we shall see, until *Descartes' readers*.

Chapter V. Several unknowns

The last strand in our cord is the appeal to several algebraic unknowns. Here, "several algebraic unknowns" are to be distinguished from a plurality of unknown quantities like the possessions of three men which are then expressed in terms of a single algebraic unknown. Unknowns become "algebraic" when they are submitted directly to algebraic operations without the mediation of other entities.

Several unknowns in Arabic and post-Arabic algebra

The main topic of the present chapter is the appearance of several unknowns in abbacus and cossic algebra. A useful background, however, is the wielding of several unknown in Arabic algebra and in two Latin works that depend directly on it: the *Liber mahameleth* and the *Liber abbaci*.

Arabic use

The Arabic use of coin names as names for extra algebraic unknowns should be familiar even to those historians of mathematics who read only European languages. Heinrich Suter showed the system in use in Abū Kāmil's *Kitāb al-Tayr*, "Book on Fowls" in [1910], though only his footnotes explain that the *x*, *y*, *z* and *u* of his translation correspond to *šai'* ("thing"), *dīnār*, *fals* and *khātam*. Recently the same short treatise has been edited and translated into French by Roshdi Rashed [2012: 731–761]. That volume also contains an edition and translation of Abū Kāmil's *Algebra*, where similarly several unknowns are used repeatedly (pp. 368–371, 394–397, 400*f*, 406*f*, 408–411, 430*f*, 654–677).

We may look at the beginning of two of these examples. Pp. 396*f* proposes an alternative solution to the problem (I translate from Rashed's French)

> You divide ten into two parts, you divide the large by the smaller, then you add the quotient to ten, then you multiply the sum by the smaller part, and one has forty-six dirhams.

A first solution identifies the smaller part with a *thing*, and obtains an equation with a single unknown. Then the alternative,

> posit that one of the things be a *thing* and the other ten less a *thing*, and then you say, we have divided a *thing* by ten less a *thing*, and one gets a *dīnar*. And we have divided ten less a *thing* by a *thing*, and one gets a *fals*. We add them, one has a *dīnar* plus a *fals*.
> ... .

As we see, *dīnar* and *fals* are introduced here as auxiliary unknowns. They facilitate the formulation of the argument but are not strictly necessary.

© The Author(s), under exclusive license to Springer Nature Switzerland AG 2024   73
J. Høyrup, *Explorations and False Trails*, SpringerBriefs in History of Science and Technology, https://doi.org/10.1007/978-3-031-48158-1_5

In the indeterminate fowl-problems, in contrast, the plurality of unknowns is primary. The first (pp. 736*f*) begins:

> If one pays to you hundred dirhams, and says to you: buy with this hundred fowl of three kinds, ducks, chickens and sparrows. Each of the ducks is five, twenty sparrows are a dirham, and the chickens are each a dirham.
>   One reasons like this: One takes the [number of] ducks a *thing*, five *things* of dirhams; the [number of] sparrows a *dīnar*, half a tenth of *dīnar* of dirhams. Some dirhams remain, hundred dirhams less five *things* less the half of a tenth of a *dīnar* [of dirhams]. ... .

But Abū Kāmil may also use other names. This happens in another divided-ten problem in the *Algebra* (p. 410). In letter formalism:

$$10 = a+b \ , \quad (^{a}/_{b}+10)\cdot(^{b}/_{a}+10) = 122^{2}/_{3} \ .$$

Abū Kāmil posits $^{a}/_{b}$ to be a "*large thing*" (presupposing $a > b$), and $^{b}/_{a}$ to be a "small *thing*". Then (R stands for the "*large thing*", r for the "*small thing*"),

$$(R+10)\cdot(r+10) = 122^{2}/_{3} \ ,$$

whence, since $rR = 1$,

$$1+10\cdot(R+r)+100 = 122^{2}/_{3} \ ,$$

from which follows

$$R+r = 2^{1}/_{6} \ .$$

Thereby, the problem is reduced to

$$10 = a+b \ , \quad ^{a}/_{b}+^{b}/_{a} = 2^{1}/_{6} \ ,$$

already dealt with by Abū Kāmil.

Abū Kāmil feels no need to explain this appeal to several unknowns.[109] He introduces nothing new or unfamiliar.

Abū Kāmil was not the only Arabic algebraic author who invented names of extra unknowns freely. In one problem, al-Karajī [ed., trans. Woepcke 1853: 141] uses *šai'* ("thing") and *qasm* ("part"). In another one (p. 139), his unknowns are *šai'* and *qist*, "share"/"measure".; even *qist* can thus have given rise to a translation "part". As we see, here at least al-Karajī does not follow the habit to use coin names for unknowns beyond the *thing*.

Around the mid-13th century, "a portion" is used as the second unknown by the Iranian jurist-mathematician al-Zanjānī in a hundred-fowl problem, see [Sammarchi 2019: 52]. The Arabic term is not mentioned but might be either *qasm* or *qist*. The problem is not

---

[109] The only explanation found is on p. 662*f*, where Abū Kāmil explains to use the term "kind" as a common name for *dirham*, *dīnar* or *fals* or other kind. Evidently this gives a name to a category whose members are already supposed to be familiar singly.

borrowed from al-Karajī, but al-Zanjānī knew al-Karajī's algebraic writings and may have been inspired.

These are the examples I know from the few Arabic algebraic works that are accessible in European languages, either translated or in paraphrase. There will certainly be many more.

Liber mahameleth

The *Liber mahameleth* is known from a translation prepared by Domingo Gundisalvi or in his environment in Toledo around 1160 – in any case a free translation, in Gundisalvi' style. The original was probably made in al-Andalus a few decades before,[110] but that is unimportant for the present point. Here we consider what was made available in Latin.

Internal references show that the treatise originally contained as systematic presentation of algebra, but that part of the work is absent from all manuscripts and may never have been translated. Many problems, however, are solved *secundum algebra*, and two of these [ed. Sesiano 2014: 258, 259*f*] operate with the two unknowns *res* and *dragma*.

Usually, as we know, *dragma* was used as the "unit" for pure numbers, but that is not the case here (the monetary unit used is *nummus*, "coin"). For readers, however, this may have presented a difficulty. Matters will not have been facilitated by the formulations of the problems, both of which deal with a purchase "of two different things", of which one is then identified in the algebraic solution with the *thing* and the other with the *dragma*. In any case I have noticed no evidence that any reader was ever inspired by these problems, even though the treatise as a whole had a modest impact. What was supposed to go by itself by Abū Kāmil was not obvious to readers of the *Liber mahameleth*.

Fibonacci

The *Liber mahameleth* and Fibonacci's *Liber abbaci* are parallel endeavours, both presenting practical arithmetic "from a higher vantage point" (to speak with Felix Klein). It is next to certain, however, that Fibonacci did not know about the *Liber mahameleth*, although there is strong evidence that he used other material already translated from the Arabic into Latin and probably translated in the Iberian Peninsula [Høyrup 2021b, *passim*].

Fibonacci's way to use several algebraic unknowns is also wholly different from what we see in the *Liber mahameleth*. Most instances occur in the *Liber abbaci*, but one is found in the *Flos*.

The latter [ed. Boncompagni 1862: 236][111] (a pure-number version of the classical recreational riddle about buying a horse) runs as follows (emphasis added in order to facilitate reading):[112]

---

[110] See [Høyrup 2021a: 42–44].

[111] Already observed by Vogel [1971: 610].

[112] In a similar problem (though with only four participants), Abū Kāmil makes use of four unknowns

five numbers, of which the *first with the half* of the second and third and fourth makes as much as the *second with the third part* of the third and fourth and fifth numbers, and as much as the *third with the fourth part* of the fourth and the fifth and the first numbers, and also as much as the *fourth with the fifth part* of the fifth and the first and the second numbers, and besides as much as the *fifth number with the sixth part* of the first and the second and the third numbers.

In symbolic abbreviation:

$$A + \tfrac{1}{2}(B+C+D) = B + \tfrac{1}{3}(C+D+E) = C + \tfrac{1}{4}(D+E+A) =$$
$$D + \tfrac{1}{5}(E+A+B) = E + \tfrac{1}{6}(A+B+C).$$

It would not be easy to solve this indeterminate problem without some kind of algebraic reflection or calculation. Fibonacci goes on:

> In order to find this, I thus posited[113] for the first number *causa*,[114] for the fifth *thing*, and for the number to which they are equal under the given conditions, I randomly posited 17.

After protracted arguments and reduction (almost 700 words), this yields two equations:

$$thing = (3 - \tfrac{1}{33})causa + 3^{20}\!/_{33}$$

and

$$thing + {}^{8}\!/_{15}\,causa = 15^{13}\!/_{15}.$$

Inserting the former into the latter and multiplying by 165 Fibonacci finds that

$$578\,causa = 2023$$

whence

$$causa = 3\tfrac{1}{2}.$$

Preferring integers, and knowing that the problem is indeterminate (though not saying that it is), Fibonacci instead chooses *causa* = *A* = 7, and derives with further intricate and somewhat elliptic arguments that *B* will then be 10, *C* will be 19, *D* will be 25, and *E* will be 29.

[ed. trans. Rashed 2012: 654*ff*].

[113] The *Flos* reports how Fibonacci solved problems with which he had been confronted, which explains this first-person singular perfect (*posui*).

[114] In medieval Latin, *causa* had come to sometimes mean an "object" or "movable thing", whence Italian *cosa* and French *chose* for "thing". Fibonacci is likely to have taken the term from medieval Catalan or Castilian, cf. [Costa & Terrés 2001: 41] and [Corominas & Paqual 1980: I, 928]. Provençal is also a possibility, cf. [Raynouard 1838: I, 358].

The use of coin names for unknowns turns up once in the *Liber abbaci*, namely in the collection of algebraic problems in chapter 15 [ed. Giusti 2020: 658–661]. In an alternative solution to this divided-ten problem

> I divided 10 in two parts, and divided the larger by the smaller, and the smaller by the larger; and aggregated that which resulted from the division, and they were 5 *denarii*

Fibonacci suggests that

> you posit one of the two parts a *thing*, and the other certainly 10 less a *thing*. And let from the division of 10 less a *thing* in a thing a *denarius* result.

Unfortunately, Fibonacci also uses *denarius* as the "unit" for pure numbers, and the calculations become meaningless.[115] I have analyzed to procedure in detail in [2019b: 32–35], and there is no need to repeat. It is obvious, however, that Fibonacci copies from as source and does not know the habit to use coin names for unknowns. The problem belongs within a whole cluster borrowed from an already existing Latin translation of a treatise descending from Abū Kāmil's *Algebra*, and mostly Fibonacci understands it so well that he can change its approach when he does not like it [Høyrup 2021b: 35–38; 2022: 180–183]. On this point, however, he does not understand and tries to make the best of it.

More interesting is what he does in a "purse-finding problem" in Chapter 12 [ed. Giusti 2020: 355]:

> Two men, who have *denari*, find a purse containing *denari*. When they have found it, the first says to the second, "if I get the *denari* in the purse together with the *denari* I have, then I shall have three times as much as you". Against which the other answered, "and if I get the *denari* of the purse together with my *denari*, I shall have four times as much as you".

A first solution goes through some arithmetical transformations and then applies a single false position. An alternative solution is made by *regula recta* (above, p. 31) (not identified by name here). The possession of the first man is posited to be a *thing*, and then Fibonacci operates with the *thing* and the *purse* (*bursa*) on an equal footing. Since *thing*+*purse* is thrice B, B must be $\frac{1}{3}(thing+purse)$. Therefore, if the second man gets the purse, he will have $purse+\frac{1}{3}purse+\frac{1}{3}thing$, which will be 4*thing*. Therefore 4*purse* = 11*thing*. In consequence, $p : A = 11 : 4$.

Since the purse conserves its name while changing its role, one needs to read attentively in order to discover that two *algebraic* unknowns are in play.

---

[115] Giusti, presupposing that correct calculations are intended, corrects the text. His critical apparatus as well as Baldassare Boncompagni's edition of a single manuscript [1857: 435*f*] shows what is really in the text.

Another instance [ed. Giusti 2020: 426] turns up within a sequence of problems about composite travels. The first of these [ed. Giusti 2020: 417] runs like this:

> Somebody proceeding to Lucca made double there, and disbursed 12 δ [*denarii*]. Going out from there he went on to Florence; and made double there, and disbursed 12 δ. As he got back to Pisa, and doubled there, and disbursed 12 δ, nothing is said to remain for him. It is asked how much he had in the beginning.

This could be solved step by step backwards, as mostly done: Before disbursing 12 δ in Pisa, he has 12 δ, that is, coming to Pisa he must have 6 δ, which have been left over in Florence after he disbursed 12 δ there. Before disbursing 12 δ in Florence he therefore had 18 δ, and coming to Florence hence 9 δ. Etc.

Fibonacci instead makes the tacit false position that the initial capital is 1 δ. He prescribes a sequence of unexplained numerical steps, whose underlying explanation is this: Without disbursements, the initial 1 δ would grow to a "Pisa value" of 8 δ. Actually, however, it grows to equal the Pisa value of the disbursements, which is $(2 \cdot 2 + 2 + 1) \cdot 12$ δ = 84 δ.

The following problems are more complex, but the basic idea underlying the solutions remains the same.

Yet for this problem [ed. Giusti 2020: 426] that will not do:

> Again, in a first travel somebody made double; in the second, of two, three; in the third, of three, 4; in the fourth, of 4, 5. And in the first travel he expended I do not known how much; in the second, he expended 3 more than in the first; in the third, 2 more than in the second; in the fourth, 2 more than in the third; and it is said that in the end nothing remained for him. And let the expenditures and his capital be given in integers. We therefore posit by *regula recta* that his capital was an *amount* [*summa*], and the first expenditure a *thing*.

This time Fibonacci makes the calculation stepwise, positing explicitly *amount* and *thing* as algebraic unknowns. We observe that Fibonacci knows the problem to be indeterminate, which allows him to ask for a solution in integers.

After the first travel, our merchant is seen to possess $2amount - thing$; after the second, he has $3amount - 2\frac{1}{2}thing - 3\delta$; etc. In this way we end up with the rhetorical equation

$$5amount - 6\frac{5}{12}thing - 18\frac{1}{4}\delta = 0$$

or, "if all-over $6\frac{5}{12}thing$ and $18\frac{1}{4}\delta$ are added",

$$5amount = 6\frac{5}{12}thing + 18\frac{1}{4}\delta$$

with the request that *amount* and *thing* have to be integers. With a clever stepwise procedure Fibonacci finds as possible solution the *amount* to be 46, and the *thing* to be 33. Since the equation can be transformed into $60amount = 77thing + 219\delta$, other solutions are found by adding as many times as you will 60 to the first expenditure, that is, to 33, and as many times 77 to the capital that was found, that is to 46, and you will have what was asked for in ways without end.

In a variant of the problem the traveller is left in the end with a net profit of 12 δ, in total thus with the initial capital and 12 δ. Here Fibonacci applies the *regula versa*, starting the construction of the equation from the final instead of the initial situation but using the same two unknowns.

A last instance of interest turns up in the alternative solution "according to the investigation of proportions" of a problem about three men having *denarii* [ed. Giusti 2020: 529],

the first asks the last two for $\frac{1}{3}$ [of what they have], and states that then he shall have 14; the second asks the third for $\frac{1}{4}$ of his *denarii*, and says he shall then have 17 *denarii;* the third, indeed, asks the first for $\frac{1}{5}$ of his *denarii*, and says he shall have 19 *denarii*.

The alternative solution [ed. Giusti 2020: 530*f*] asks to

posit that the second and the third man have a *thing*. Therefore the first has 14, less a third of a *thing*. Then posit that the third has a *part* of a thing. Therefore the second has a *thing*, less a *part*.

This gives the equations

$$^{11}/_{12}thing - ^3/_4part = 13\frac{1}{2} \; , \quad ^4/_5part + ^2/_{15}thing = 16\frac{1}{5} \; ,$$

of which the latter after multiplication by $\frac{5}{6}$ becomes

$$^2/_3part + ^1/_9thing = 13\frac{1}{2} \; .$$

The right-hand side being equal, the ratio $r : p$ can be determined, whence also the ratio $r - p : p$. This leads to the solution.

The name *"part of a thing"* looks strange but might be borrowed from the Arabic tradition – it is suspiciously close to what was reported above (p. 74) from al-Karajī and al-Zanjānī.

As we shall see below (p. 98), there are traces in Fibonacci's text that he made secret use of some unidentified tool (probably line diagrams) which allowed him to work algebraically with many unknowns. However, since nobody guessed at the time, this is not pertinent to what we are dealing with here.

Abbacus occurrences of several unknowns around 1400

Two almost contemporary abbacus treatises that were discussed in earlier chapters make use of two unknowns. Their ways are completely different, and none of them draw directly on Fibonacci or on anything Arabic I can identify.

Antonio de' Mazzinghi

Antonio's *Fioretti* were mentioned above (p. 27), and it was pointed out that what we have is a philologically conscientious copy of a work in progress.[116] This allows us to follow how Antonio gradually develops a technique for working with several unknowns, and to conclude that his idea is original and not borrowed. His use is also original, and not related to what Fibonacci or other predecessors had done.

The problems are numbered in the manuscript – almost certainly by Antonio himself. I shall use this numbering for references in the following discussion.

The first step toward two unknowns is taken in #9. Here, two numbers are asked for – for brevity A and B – such that

$$AB = 8 \ , \quad A^2 + B^2 = 27 \ .$$

A first solution [ed. Arrighi 1967a: 28], "though the case does not come in discrete quantity", makes use of *Elements* II.4, according to which (when it is read as dealing with "quantities" and not line segments)

$$A^2 + B^2 + 2AB = (A + B)^2 \ .$$

This leads to

$$A = \sqrt{10\tfrac{3}{4}} + \sqrt{2\tfrac{3}{4}} \ , \quad B = \sqrt{10\tfrac{3}{4}} - \sqrt{2\tfrac{3}{4}} \ ,$$

and also tells us that Antonio's use of "quantity" has nothing to do with that of Aristotelian or scholastic philosophy (where it would refer to lengths, weights and other continuous magnitudes, and be opposed to numbers). A "quantity", for Antonio, is a number or, when needed (as here), an expression involving radicals.

Next Antonio explains that

> we can also make it by the equations [*aguagliamenti*] of algebra; and that is that we posit that the first quantity[117] is a *thing* less the root of some quantity, and the other is a *thing* plus the root of some quantity. Now you will multiply the first quantity [A] by itself and the second quantity [B] by itself, and you will join together, and you will have 2 *censi* and an unknown quantity, which unknown quantity is that which there is from 2 *censi* until 27, which is 27 less 2 *censi*, where the multiplication of these quantities [those of which the square root was taken] is $13\tfrac{1}{2}$ less a *censo*. The smaller part is thus a *thing* minus the root of $13\tfrac{1}{2}$ less a *censo*, and the other is a *thing* plus the root of $13\tfrac{1}{2}$ less 1 *censo*. [...].

---

[116] Fols 451^r–474^v in Benedetto's *Praticha*. The transcription published by Arrighi [1967a] is convenient but should be controlled against the manuscript.

[117] The two numbers of the statement have now become "quantities" – Antonio often replaces one word by the other. As we shall see in the following lines, that creates some confusion, only to be kept under control by keen unspoken awareness of what the various "quantities" refer to. Further on, however, Antonio turns out to be aware of the difficulty and to know how to circumvent it.

A procedure using the two algebraic unknowns *thing* and (some) *quantity* (say, q) would have started with these steps (C stands for *censo*):

$$A = t + \sqrt{q} \ , \quad B = t - \sqrt{q}$$

$$A^2 + B^2 = 2C + 2(\sqrt{q})^2 = 2C + 2q$$

whence

$$q = 13\tfrac{1}{2} - C \ .$$

This corresponds precisely to the numerical steps in Antonio's argument, and obviously to his understanding. But what he does can instead be expressed

$$a = t + \sqrt{?} \ , \quad b = t - \sqrt{?}$$
$$a^2 + b^2 = 2C + ?? \ ,$$

and the fact that "??" equals two times "?" remains private knowledge.

From this point onward, the method is algebraic, but with only one unknown (and the procedure is impeccable).

In the next problem (#10) [ed. Arrighi 1967a: 30] we read:

> Find two numbers whose squares are 100, and the multiplication of one by the other is 5 less than the squared difference. Posit that the first number be a *thing* plus the root of some quantity, and the second be a *thing* less the root of some quantity, and multiply each number by itself and join the squares, they make two *censi* and something not known. And these squares should make up 100. Whence this unknown something is the difference there is from 100 to 2 *censi*, which is 100 less 2 *censi*. [...].

Antonio here comes closer but still does not fully implement the possibility of working *algebraically* with two unknowns. He is preparing mentally, however, and in problem #18 [ed. Arrighi 1967a: 41] the idea is unfolded:

> Find two numbers which, one multiplied with the other, make as much as the difference squared, and then, when one is divided by the other and the other by the one and these are joined together make as much as these numbers joined together. Posit the first number to be a *quantity* less a *thing*, and posit that the second be the same *quantity* plus a *thing*. Now it is up to us to find what this *quantity* may be, which we will do in this way. We say that one part in the other make as much as to multiply the difference there is from one part to the other in itself. And to multiply the difference there is from one part to the other in itself makes 4 *censi* because the difference there is from a *quantity* plus a *thing* to a *quantity* less a *thing* is 2 *things*, and 2 *things* multiplied in itself make 4 *censi*. Now if you multiply a *quantity* less a *thing* by a *quantity* plus a *thing* they make the square of this *quantity* less a *censo*; so the square of this *quantity* is 5 *censi*. And if the square of this *quantity* is 5 *censi*, then the *quantity* is the root of 5 *censi*; whence we have made clear that this *quantity* is the root of 5 *censi*. And therefore the first number was the root of 5 *censi* less a *thing* and the second number was the root of 5 *censi* plus a *thing*. We have thus found 2 numbers which, one multiplied in the other, make as much as to multiply

the difference of the said numbers in itself; and one is the root of 5 *censi* less a *thing*, the other is the root of 5 *censi* plus a *thing*. Now remains for us to see whether one divided by the other and the other by the one and these two results joined together make as much as the said numbers. Where you will divide the root of 5 *censi* less a *thing* by the root of 5 *censi* plus a *thing*, this results, that is, $\frac{r.\ of\ 5\ C\ less\ 1\rho}{r.\ of\ 5\ C\ plus\ 1\rho}$. And then you will divide the root of 5 *censi* plus 1 *thing* by the root of 5 *censi* less a *thing*, $\frac{r.\ of\ 5\ C\ plus\ 1\rho}{r.\ of\ 5\ C\ less\ 1\rho}$ results. And these two results should be joined together; where you will multiply the root of 5 *censi* plus a *thing* across,[118] that is, by the root of 5 *censi* plus *a thing*, they make *censi* plus the root of 20 *censi of censo*; and further multiply root of 5 *censi* less a *thing* across, that is, by root of 5 *censi* less a *thing*, they make 6 *censi* less root of 20 *censi of censo*. Which, joined with 6 *censi* and root of 20 *censi of censo*,[119] make 12 *censi*. And this quantity we should divide in the multiplication of the root of 5 *censi* less a *thing* in root of 5 *censi* plus a *thing*, which multiplication is 4 *censi* because root of 5 *censi* in root of 5 *censi* make 5 *censi*, and a *thing* plus multiplied in a *thing* less make a *censo* less, and when it is detracted from 5 *censi*, 4 *censi* remain, and multiplying 1 *thing* plus by root of 5 *censi* and 1 *thing* less by root of 5 *censi*, their joining makes 0. So the said multiplication, as I have said, is 4 *censi*, so these two results are 12 *censi* divided in 4 *censi*, from which comes 3. And we want they should make as much as the sum of the said numbers, whence it is needed to join the root of 5 *censi* less a *thing* with the root of 5 *censi* plus a *thing*, they make 2 times the root of 5 *censi*, which is the root of 20 *censi*. Whence the joining of the said numbers is the root of 20 *censi*, and we say that is should be 3; so 3 is equal to the root of 20 *censi*. Now multiply each part in itself, and you will have 9 to be equal to 20 *censi*; so that, when it is brought to one *censo*, you will have that the *censo* will be equal to $\frac{9}{20}$. So the *thing* is equal to the root of $\frac{9}{20}$, and if the *thing* is equal to the root of $\frac{9}{20}$, the *censo* will be worth its square, that is, $\frac{9}{20}$. So the first number, which was the root of 5 *censi* plus a *thing*, was $1\frac{1}{2}$ plus the root of $\frac{9}{20}$; and the second number, which was the root of 5 *censi* less a *thing*, was $1\frac{1}{2}$ less the root of $\frac{9}{20}$. And so is found the said two numbers [...].

This may have gone beyond what Antonio was able to do by mental implicit use of a second unknown, or at least beyond what he found it possible to convey to an imagined reader in this way. In any case he now makes the use of two unknowns explicit, and also chooses a more stringent language, pointing out that the *same* quantity is meant in the two positions. The remark that now "it is up to us to find what this *quantity* may be" shows awareness that something unfamiliar has happened – it is never explained that the *thing* has to be found, neither here nor elsewhere in problems with a single algebraic unknown.

From this point onward, *quantity* in general use (cf. note 117) disappears from all problem solutions where that term is used to designate one of two algebraic unknowns

---

[118] The cross-multiplication is shown in a symbolic operation on the two formal fractions in the margin in the manuscript (fol. 458^v).

[119] Arrighi has 20 *censi* only, but the manuscript (fol. 458^v) is correct.

(but not from other problems – in these *quantity* is still used profusely.[120]

The procedure can be translated into more familiar symbols as follows:

$$AB = (A-B)^2 , \quad ^A/_B + {}^B/_A = A+B$$

with the algebraic positions

$$A = q-t , \quad B = q+t .$$

Then

$$(A-B)^2 = 4C , \quad \text{while} \quad AB = q^2 - C ,$$

whence

$$q^2 = 5C ,$$

that is,

$$q = \sqrt{5C} .$$

In consequence we have the preliminary result

$$A = \sqrt{5C} - t , \quad B = \sqrt{5C} + t .$$

Inserting this in the other condition we get

$$\frac{A}{B} + \frac{B}{A} = \frac{\sqrt{(5C)}-t}{\sqrt{(5C)}+t} + \frac{\sqrt{(5C)}+t}{\sqrt{(5C)}-t}$$

which, after cross-multiplication, becomes

$$\frac{A}{B} + \frac{B}{A} = \frac{(\sqrt{(5C)}-t)^2 + (\sqrt{(5C)}+t)^2}{5C-C} = \frac{6C+6C}{4C} = \frac{12C}{4C} = 3 .$$

Therefore, since

$$A+B = 2q = 2\sqrt{(5C)} ,$$

$$2\sqrt{5C} = \sqrt{20\,C} = 3 ,$$

whence

$$20C = 9 .$$

Tacitly interchanging "first" and "second" number, Antonio thereby obtains that

$$B = 1\tfrac{1}{2} + \sqrt{9}/_{20} , \quad A = 1\tfrac{1}{2} - \sqrt{9}/_{20} .$$

This would probably have been very difficult even for a mathematician of Antonio's level without the explicit use of two unknowns. Once Antonio had decided to make the step,

---

[120] There are two apparent exceptions, one in the present problem ("this quantity we should divide in the multiplication of the root of 5 *censi* less a *thing* in root of 5 *censi* plus a *thing*"), one in problem 28 [ed. Arrighi 1967a: 61*f*]. Both, however, turn up after the algebraic *quantity* has been eliminated, and the problem thus reduced to one with a single unknown *thing*.

things were easy. As can be seen in marginal calculations, Antonio routinely performed formal calculations involving ρ and c or c^o (standing for *censo*).

Now, as the method has been developed and introduced, Antonio applies it even in problem #19 [ed. Arrighi 1967a: 43] although it *could* have been solved in the same way as problems #9 and #10:

Find two numbers so that the root of one multiplied by the root of the other be 20 less than the numbers joined together, and their squares joined together be 700. It is asked, which are the said numbers? You will make position that the first number be a *thing* less some quantity, and posit that the other number be a *thing* plus some quantity. And then you take the square of the first, which we said was one *thing* less one *quantity*, and its square is one *censo* and the square of this *quantity* less the multiplication of this *quantity* in a *thing*. And the square of the second number, which we say is a *thing* and some quantity, is a *censo* and the square of this quantity plus the multiplication of this quantity in a *thing*.[121] Which, joined together, make 2 *censi* and 2 squares of 2 *quantities*.[122] And we say that they should make 700, whence one of these squares is 350 less one *censo*. This quantity is thus the root of 350 less once *censo*. And we posited that the first number was one *thing* less one *quantity*, that is was hence one *thing* less the root of 350 less one *censo*. And the second number, which was posited to be a *thing* and a *quantity*, was one *thing* and root of 350 less one *censo*. And thus we have solved a part of our question, that is, to find two numbers whose squares joined together make 700. Now it remains for us to see what it makes to multiply the root of one by the root of the other. Therefore you thus have to multiply the general root of one *thing* less root of 350 less one *censo* by the general root of one *thing* plus root of 350 less one *censo*,[123] they make root of 2 *censi* less 350; and this is their multiplication. For these matters one has to keep the eye keen, I mean of the mind and the intellect, because even though they seem rather easy, none the less, who is not accustomed will err. Therefore we have thus found that this multiplication is the root of 2 *censi* less 350, and this we say is 20 less than the numbers joined together. And the said numbers joined together are 2 *things*, that is joining a *thing* less root of 350 less a *censo* with a *thing* plus root of 350 less a *censo*, which indeed make 2 *things*. Whence we have that 2 *things* less 20 are equal to the root of 2 *censi* less 350; whence, in order not to have the names of roots, multiply each part in itself, and you will have that root of 2 *censi* less 350 multiplied in itself make 2 *censi* less 350, and 2 *things* less 20 multiplied in itself make 4 *censi* and 400 less 80 *things*. So 2 *censi* less 350 are equal to 4 *censi* and 400 less 80 *things*. Where you should make equal the parts giving

---

[121] Evidently, the product of *quantity* and *thing* should be taken twice here as well as in the square of the first number. Antonio knew perfectly well how to multiply two binomials (see, for instance, the marginal calculation reproduced on p. 52). Since the "error" is repeated in subsequent problems, we may be sure that Antonio abbreviates, knowing that the two elliptical expressions cancel each other (as they also do in that calculation).

[122] *2 quadrati di 2 quantità*, meaning "the two squares coming from the two distinct quantities".

[123] Cf. above, page 52.

to each part 80 *things* and removing 2 *censi;* and we shall have that 2 *censi* and 740 are equal to 80 *things*, which is the fifth rule.[124] Where you bring to one *censo*, and you will have one *censo* and 375 equal to 40 *things*. Where you will halve the *things*, and the half be 20, multiply in itself, they make 400, detract the number, they will make 25, that is, detracting 375 from 400, of which 25 take the root, which is 5, and detract it from 25, 15 remain. And you will say that the *thing* is worth 15, and the *censo* will be worth its square, which is 225. Whence the first number, which we posited that it was a *thing* less root of 350 less a *censo*, detract 225, which is worth the *censo*, from 350, 125 remain. And you will say, one part was 15 less root of 125, and the second number was 15 plus root of 125. [...].

In our usual translation:

$$\sqrt{A}\cdot\sqrt{B} = A+B-20 \, , \ \ A^2+B^2 = 700 \, ,$$

with the position

$$A = t-q \, , \ \ B = t+q \, ,$$

where Antonio no longer feels the need to point out that the two "some quantity" (*alchuna quantità*) refers to *the same* quantity. He does not quite return to the formulation of problems 9 and 10, $A = t-\sqrt{q}$, $B = t+\sqrt{q}$, since with the explicit position of q he can now operate freely with its square. Antonio calculates

$$A^2 = C+q^2-[2]qt \, , \ \ B^2 = C+q^2+[2]qt \, ,$$

whence

$$2C+2q^2 = 700 \, , \ \ q^2 = 350-C \, , \ \ q = \sqrt{(350-C)} \, .$$

Therefore

$$A = t-\sqrt{(350-C)} \, , \ \ B = t+\sqrt{(350-C)} \, .$$

This partial answer is inserted in the other condition:

$$AB = \sqrt{t-\sqrt{(350-C)}} \cdot \sqrt{t+\sqrt{(350-C)}} = \sqrt{C-(350-C)} = \sqrt{2C-350} \, ,$$

a calculation which seems straightforward but where, according to Antonio, the untrained will none the less err.[125] We now have

$$\sqrt{2C-350} = A+B-20 = 2t-20$$

and thus after squaring

---

[124] That is, the fifth standard "case" (equation type) of abbacus *aliabra* (and al-Khwārizmī's *al-jabr*), "*censi* and number are equal to *things*". This is the case with a double solution, which Antonio neglects here – the alternative solution leads indeed to complex and thus impossible values for *a* and *b*.

[125] Those who doubt Antonio's words should have a look at the mistake described in note 102.

$$2C - 350 = 4C + 400 - 80t ,$$

which can be reduced to

$$2C + 750 = 80t .$$

Solving this equation by means of the standard rule for the fifth case Antonio finds $t = 15$ – silently discarding the other solution $t = 25$, cf. note 124.

Several other problems in the *Fioretti* are solved by means of two algebraic unknowns: #20, #21, #22 (twice during the procedure), #24, #25 and #28. All make the position

$$a = t - q , \quad b = t + q ,$$

and all *could* have been solved in the same way as number 9 and number 10, if only the position had been

$$a = t - \sqrt{?} , \quad b = t + \sqrt{?} ,$$

that is, with an implicit second unknown. Only one detail tells something of interest – namely, this passage from #20 [ed. Arrighi 1967a: 44]:

> Find two numbers so that their roots joined together make 6 and their squares be 60, that is, the joining of the squares be 60. Posit the first number to be a *thing* less the root of some quantity, that is less some quantity; the other posit to be a *thing* plus the said quantity. [...].

This confirms that Antonio as copied by Benedetto presents us with a work in progress – if the *Fioretti* had been polished, it would not have left a formulation "root of some quantity" then to be corrected. More striking: the slip shows that Antonio at first had in mind the method of problems 9 and 10; it is a plausible guess that he used an earlier solution of the problem – probably his own, nobody else in Italy between Fibonacci and Antonio is known to have possessed adequate mathematical capabilities together with similar interest.

Antonio *may* have been aware of Fibonacci's use of two unknowns – he knew at least some of his works and may well have read them in sufficient depth. However, what he develops here is quite different from what Fibonacci offers. With exception of the botched problem from chapter 15 section 3 of the *Liber abbaci*, all problems where Fibonacci uses two unknowns are linear, as the *regula recta* in general. Those of Antonio are not. Moreover, Antonio understands his problems to belong within the area of *aliabra* – his *thing* multiplied by itself becomes a *censo*. The method developed by Antonio in the *Fioretti* is an independent creation.

It remains a possible hypothesis that the name *quantità* for the second unknown was a borrowing; we do not know it from any earlier abbacus writer, but as we shall see below (p. 105), more went on before Pacioli's times with two unknowns than we know about; maybe also before Antonio's times. However, Antonio's groping start with "some quantity" before he comes to "quantity" speaks against the hypothesis.

The Florentine *Tratato*

The Florentine *Tratato sopra l'arte dell arismeticha* (above, p. 28), only slightly later than Antonio, approaches the use of several algebraic unknowns in a wholly different way. Its author (we seem to be confronted with an original composition) is a brilliant algebraist, as shown by his transformation of cubic equations – see [Høyrup 2019a: 331*f*]. This makes this approach all the more striking.

The treatise contains a large collection of problems illustrating the 22 standard rules (Jacopo's 20 rules, plus the missing two biquadratics – those problems of the fourth or lower degree that could be solved correctly by the methods at hand). Then come, in the very end, four problems of a different character.

Two of these problems are of type "finding a purse", two "purchase of a horse". All four make use of two algebraic unknowns (partial use, as we shall see). None of them take note of that, in spite of being provided with a metamathematical commentary (here in spaced writing). At first we have a purchase, not of a horse but of a goose:[126]

> Three have *denari* and they want to buy a goose, and none of them has so many *denari* that he is able to buy it on his own. Now the first says to the other two, if each of you would give me $\frac{1}{3}$ of his *denari*, I shall buy the goose. The second says to the other two, if you give me $\frac{1}{4}$ plus 4 of your *denari* I shall buy the goose. The third says to the other two, if you give me $\frac{1}{4}$ less 5 of your *denari* I shall buy the goose. Then they joined together the *denari* all three had together and put on top the worth of the goose, and the sum will make 176, it is asked how much each one had for himself, and how much the goose was worth. A c t u a l l y I b e l i e v e t o h a v e s t a t e d s i m i l a r q u e s t i o n s a b o u t m e n in the treatise,[127] but wanting to solve certain questions in a new way I have found new cases which I do not believe to have (already) treated. [...]. Therefore I have made it in such way that in this one and those that follow it will have to be shown that the question examined by the thing will lead to new questions that cannot be decided without false position. [...]. I shall make this beginning, let us make the position that the first man alone had a *thing*, whence, made the position, you shall say thus, if the first who has a *thing* asks the other two so many of their *denari* that he says to be able to buy the goose, these two must give to the first that which a *goose* is worth less what a *thing* is worth, which the first has on his own. So that the first can say to ask from the other two a *goose* less a *thing*, and you know that the first when he asks for the help of the others asks for $\frac{1}{3}$ of their *denari*. So the two without the first must have so much that $\frac{1}{3}$ of their *denari* be a goose less a *thing*, and in this way you see clearly that the second and the third together have 3 *geese* less 3 *things*. Now it is to be seen what

---

[126] I translate from [Franci & Pancanti 1988:145–150], correcting a few mistakes in agreement with the manuscript.

[127] Namely in the sense that fols 97ᵛ–110ʳ contain a large number of "give and take", "purchase of a horse" and "finding a purse" problems.

all the three have, and it is clear that the first by himself has a *thing* and the other two have 3 *geese* less 3 *things*, so that all three have 3 *geese* less 2 *things*. Now we must come to the second, who asks from the other two $\frac{1}{4}$ plus 4 of their *denari* and says to buy a *goose*. I say that when the second has had as help of the other two the part asked for, he shall find to have a *goose*).

Protracted arguments lead to conclusion that B is $\frac{1}{3}$ *goose* plus $\frac{2}{3}$ *things* less $5\frac{1}{3}$ in number (A, B and C being the three original possessions). Since $B+C$ has been seen to be 3 *geese* less 3 *things*, C is $2\frac{2}{3}$ *geese* and $5\frac{1}{3}$ in number less $3\frac{1}{2}$ *things*. Using then that $C+\frac{1}{4}(A+B)-5$ is a *goose*, it is found (I skip intermediate steps) that $1\frac{3}{4}$ *geese* equals $3\frac{1}{4}$ *things* and 1 in number, or, multiplying "in order to eliminate fractions",

$$7geese = 13things + 4 \ .$$

Moreover, since $A+B+C$ was seen to equal 3 *geese* less 2 *things*, and these together with the *goose* equalled 176

$$4geese - 2things = 176 \ .$$

Now, for instance, the *thing* might be found from the latter equation (namely, to be 2 *geese* less 88) and be inserted in the former; that would easily lead to the goal, and that is what Fibonacci or Abū Kāmil would have done. Instead the author goes on with an equally protracted solution of the two equations by means of a double false position – a method that is already opaque in itself in the sense that it is never explained why it works,[128] but more opaque here since two equations are combined.[129]

The other three problems are similar. To my knowledge, their way is unique in the history of mathematics. The style – taking the *goose* as an unknown that can be added, subtracted and multiplied by a coefficient – is too similar to what we find earlier in the Fibonacci problems and later in Benedetto's two treatises to be an independent invention. It seems that the author has borrowed an idea in rarefied circulation – so rarefied that he only grasps half of it; and that he has completed it in his own way, drawing on a familiar technique.

---

[128] The method makes use of the "alligation principle": combining two wrong guesses in such a way that the errors cancel, in the same way as two alloys of different fineness are combined to yield an alloy of the fineness asked for.

[129] A partial translation of this solution can be found in [Høyrup 2019b: 55*f*]. An analysis in modern symbolism is offered by Franci and Marisa Pancanti [1988: xxiii–xxiv].

The Florentine abbacus encyclopedias

The *regula recta* (also with several unknowns) returns in the Florentine abbacus encyclopedias (above, p. 21), though now under the name *modo recto* and with unknown (or primary unknown) *quantità*. The Ottoboniano *Praticha* appears to inform us about what was currently done and known; Benedetto's *Praticha* reveals how much more could be made on this basis. Both show how easily innovations can be forgotten if nobody is interested – in manuscript culture but not only. *Perhaps* (as in the present case) to be rediscovered or reinvented independently once the context fits.

The Ottoboniano *Praticha*

On fol. 28^v, the Ottoboniano *Praticha* explains, we remember from p. 32, that the *modo recto* is used by "Leonardo [Fibonacci] and all the others who understand". The writer thus knows the technique from the *Liber abbaci*. The reference to "all the others who understand" shows, however, that he also knows if from a general abbacus tradition, within which, as he says, "some say it is one of the exemplary modes of algebra". The use of *quantità* as unknown and the reference to a *modo* shows that the writer's main reference is the living abbacus tradition, not Fibonacci.

Many problem solutions make use of the technique, often accompanied by symbolic calculations in the margin where the *quantità* is abbreviated *q*. Others do not speak explicitly about it, but there is obvious continuity in the way the marginal calculations (and the solutions in general) proceed. One (fol. 132^{r-v}) is of particular interest:

> 5 eggs and 4 oranges and 10δ are worth 8 eggs and 2 oranges and 6 δ. And 7 eggs and 6 oranges less 3 δ are worth 5 eggs, 4 oranges and 7 δ. It is asked, what is an egg worth, and what is an orange worth? This case has been given to me a few days ago to solve.

The final clause shows that this was the level that was considered difficult, and reversely the purpose of such problems – only difficult matters could serve as challenges.

The marginal calculations are algebraic, though using no symbols for the two prices but positions in columns; the question is already formulated as a set of two rhetorical equations. We may provide the columns with the headings which the writer kept in his mind, in agreement with what is found in the text. As also elsewhere in the three encyclopedias (above, p. 54), the long stroke serves as equation sign:

e	o	δ		e	o	δ
5	4	10		8	2	5
		5				
5	4	5	———	8	2	–
	4	5		3	2	
	2					
	2	5		3		

The equation $5e+4o+10\delta = 8e+2o+5\delta$ is thus reduced in steps to $2o+5\delta = 3e$ (e and o being indicated by column positions). In the same way, $7e+6o-3\delta = 5e+4o+7\delta$ is next reduced to $2e+2o = 0o+10$ by means of a similar scheme.

Now the *modo retto* is made use of, and the worth of the orange is posited to be a *quantity*. Therefore 3 eggs are worth 2 *quantities* and 5δ; 1 egg hence $\frac{2}{3}$ *quantity* and $1\frac{2}{3}\delta$; and 2 eggs and 2 oranges in consequence $3\frac{1}{3}$ *quantity* and $3\frac{1}{3}\delta$, but also 10δ. Therefore, $3\frac{1}{3}q$ are worth $6\frac{2}{3}\delta$; the *quantity* hence equals 2δ, which is thus the worth of an orange, while the egg, being worth $\frac{2}{3}$ *quantity* and $1\frac{2}{3}\delta$, is worth 3δ.

Even in the absence of explicit symbols, the techniques of handling two linear equations with two unknowns could thus be implemented without being seen as a striking innovation.

Benedetto

Benedetto goes far beyond this informal approach and transforms it stepwise into a genuine technique (used with up to five unknowns). The main (and only known complete) manuscript of his *Praticha* is an autograph, sometimes copying explicitly from earlier work (as we have seen, al-Khwārizmī, Fibonacci, Biagio, Antonio, but also others). The part under discussion here, however, is a work in progress, which allows us to see how the idea develops.

The first step is taken on from fol. 270[v] onward:[130]

(fol. 270v)Four have *denari*, and walking on a road they found a purse with *denari*. The first and second say to the third, if you give us the purse we shall have 2 times as much as

---

[130] Benedetto's grammar is less perfect than his mathematics; I shall not try to improve his style. I omit what Benedetto has deleted. Words in ⟨ ⟩ have been forgotten by Benedetto; they are restored from his preceding symbolic calculations and after further control that they are presupposed in what follows. Errors Benedetto has overlooked are conserved but pointed out in notes – they as well as all other corrections all concern the secondary description in words, not the original symbolic calculations. "First", "second", "third", "fourth" and "purse" are italicized when they serve as algebraic unknowns.

A transcription of those of Benedetto's Tuscan texts that are translated here can be found in [Høyrup 2020: 75–80].

you. The second and third men say to the fourth, if we had the *denari* of the purse we should have 3 times as much as you. The third and fourth say to the first, if we had the *denari* of the purse we should have 4 times as much as you. The fourth and the first say to the second, if you give us the *denari* of the purse we shall have 5 times as much as you. It is asked how much each one had, and how many *denari* there were in the purse. We shall do in this way, you shall say, the *first* and the *second* with the *denari* of the purse say to have 2 times as much as the *third* man. Whence the *third* man by himself had the $\frac{1}{2}$ of that which the *first* and the *second* and the *purse* have. And mark (*segnia*[131]) this. And then you shall say, the *second* and the *third* man with the *purse* have 3 times as much as the fourth, so the fourth man had the $\frac{1}{3}$ of that which the *first* and *second* have, and of the *purse*. And mark even this. And then you shall say, the *third* and the fourth man with the *purse* had 4 times as much as the *first*, and therefore the *first* man by himself had the $\frac{1}{4}$ of that which the *third* and the fourth man had, and of the *purse*. An mark this. And then you shall say, the fourth and *first* have with the *purse* 5 times as much as the *second*, so that the *second* will have the $\frac{1}{5}$ of that which the *first* and fourth man have, and of the *purse*. And this is marked. And you shall bring the *denari* of the *third* to a comparison.[132] And you shall say, the *denari* of the *third* man is as much as the $\frac{1}{2}$ of the *denari* of the *first* and the *second* and of the *purse*. From where it is to be known, how much are the $\frac{1}{2}$ of the *denari* of the *first*, which we have ⸢brought together,⸣ that the *denari* of the *first* are the $\frac{1}{4}$ of the *denari* of the *third* and fourth man and of the *purse*, and let the $\frac{1}{2}$ of the *denari* of the *first* be $\frac{1}{8}$ of the *denari* of the *third* and fourth and of the *purse*. Therefore you shall say that the *denari* of the *third* should be as much as the $\frac{1}{2}$ of the *denari* of the *second* and of the *purse* and as much as $\frac{1}{8}$ of the *denari* of the *third* and fourth and of the *purse*. Therefore you shall take away $\frac{1}{8}$ of the *denari* of the *third* and join $\frac{1}{8}$ of *purse* to $\frac{1}{2}$ *purse*. And we shall have that $\frac{7}{8}$ of the *denari* of the *third* are $\frac{1}{8}$ of the *denari* of the fourth and $\frac{1}{2}$ of the *denari* of the *second* and $\frac{5}{8}$ of *purse*. And then (fol. 271r)it is to be seen how much $\frac{1}{8}$ of the *denari* of the fourth are, where we say that all the *denari* of the fourth man are $\frac{1}{3}$ of the *second* and the *third* man and of the *purse*. Therefore $\frac{1}{8}$ of the *denari* of the *third* man will be $\frac{1}{24}$ of the *denari* of the *second* and of the *third* and of the *purse*. Therefore you shall take away from $\frac{7}{8}$ of the *denari* of the *third* $\frac{1}{24}$ of the denari of the *third* man and above $\frac{5}{8}$ of *purse* you shall put $\frac{1}{24}$ of *purse*, and above $\frac{1}{2}$ of the *second* you shall put $\frac{1}{24}$ of the *denari* of the *second*, and we shall have that $\frac{5}{6}$ of the *denari* of the *third* man are as much as the $\frac{2}{3}$ of the *purse* and $\frac{13}{24}$ of the *denari* of the *second*. And you shall say, if $\frac{5}{6}$ of the *denari* of the *third* man are as much as the $\frac{2}{3}$ of the *purse* and $\frac{13}{24}$ of the *second*, how much will all the *denari* of the *third* be? Where you shall divide $\frac{2}{3}$ of the *denari* of the *purse* and $\frac{13}{24}$ of the *denari* of the *second* in $\frac{5}{6}$, from which comes $\frac{4}{5}$ of the *denari* of the *purse* and $\frac{13}{20}$ of the *denari* of the *second*. This is taken note of (*notato*[133]). And in the same way you shall make

---

[131] Technically, this can be seen to mean that it is written down as a symbolic equation. Cf. note 133.

[132] Technically, as we see in the following, this refers to an algebraic substitution.

[133] The verb *notare* may mean both "write down" and "consider". "Take note" should be similarly

the *denari* of the fourth. It is true that it could be done without that, but since it is in the castelet[134] this order shall be pursued. You shall say, the fourth has the $\frac{1}{3}$ of the *denari* of the *second* and the *third* and of the *purse*. Where you shall get the $\frac{1}{3}$ of the *denari* of the *second* without the others, and you shall say, the *second* has as much as the $\frac{1}{5}$ of the *first* and of the fourth and of the *purse*. Wherefore the $\frac{1}{3}$ of the *denari* of the *second* will be $\frac{1}{15}$ of the *denari* of the *first* and of the fourth and of the *purse*. Where from the *denari* of the *first* you shall take away $\frac{1}{15}$ and you shall join to $\frac{1}{3}$ of *purse* $\frac{1}{15}$ of *purse*, and we shall have that $\frac{14}{15}$ of the fourth are as much as $\frac{1}{15}$ of the *first* and $\frac{1}{3}$ of the *third* and $\frac{2}{5}$ of *purse*. Then $\frac{1}{15}$ of the *first* will be $\frac{1}{60}$ of the *third* and $\frac{1}{60}$ of the fourth and of the *purse*, where from $\frac{14}{15}$ of the *denari* of the fourth take away $\frac{1}{60}$, and above $\frac{1}{3}$ of the *denari* of the *third* man join $\frac{1}{60}$ of the *denari* of the *third* man, and above the $\frac{2}{5}$ of the *denari* of the *purse* put $\frac{1}{60}$ of *purse*, and you shall have that $\frac{11}{12}$ of the *denari* of the fourth are $\frac{7}{20}$ of the *denari* of the *third* and $\frac{5}{12}$ of the *denari* of the *purse*. And you shall say, if $\frac{11}{12}$ of the fourth are $\frac{7}{20}$ of the *denari* of the *third* man and $\langle\frac{5}{12}\rangle$ of the *purse*, what will all the *denari* of the fourth be? Where you divide by $\frac{11}{12}$, from which comes $\frac{21}{55}$ of the *third* and $\frac{5}{11}$ of the *purse*, and as much has the fourth man. And this is done. And we shall make position that the *second* has a *thing*, that is, a *quantity*.[135] And because we have found that the *first* man has the $\frac{4}{5}$ of the *denari* of the *purse* and the $\frac{13}{20}$ of the *denari* of the *second*, the *third* man will thus have $\frac{13}{20}$ of *quantity* and $\frac{4}{5}$ of *purse*. And the fourth man, whom we have found that he has the $\frac{21}{55}$ of the *third* and $\frac{5}{11}$ of the *purse*. First you shall take the $\frac{21}{55}$ of that which the *third* has, of, that is, $\frac{13}{20}$ of *quantity* and $\frac{4}{5}$ of *purse*, which are $\frac{273}{1100}$ of *quantity* and $\frac{84}{275}$ of *purse*, and to the $\frac{84}{275}$ of *purse* you join $\frac{5}{11}$ of *purse*, they make $\frac{19}{25}$ of *purse*. And you shall say that the fourth man has the $\frac{273}{1100}$ of *quantity* and $\frac{19}{25}$ of *purse*. Now so as to know that which the *first* has you shall keep this way. The fourth and *first* with the *purse* have 5 times as much as the *second*. Therefore, if the *second* has 1 *quantity*, they will have with the *purse* 5 *quantities*. And we say that the fourth man has $\frac{273}{1100}$ of *quantity* and $\frac{19}{25}$ of *purse*. When to these has been joined a *purse* they make $\frac{273}{1100}$ $\langle$of *quantity*$\rangle$ and 1 *purse* $\frac{19}{25}$. And so much has the fourth with the *denari* of the *purse* and with the *denari* of the *first*. They should be 5 *quantities*, thus the *first* had, from the $\frac{273}{1100}$ of *quantity* and 1 *purse* $\frac{19}{25}$ until 5 *quantities*, where there are 4 *quantities* $\frac{827}{1100}$ less a *purse* $\frac{19}{25}$. And so much has the *first*. And in this way we have established that the *first* has 4 *quantities* $\frac{823}{1100}$ $\langle$less 1 *purse* $\frac{19}{25}\rangle$. The *second* has a *quantity* and the *third* has $\frac{13}{20}$ of *quantity* and $\frac{4}{5}$ of *purse*. And the fourth has $\frac{273}{1100}$

---

ambiguous. *Notare* as well as the functionally similar *segnare* are used about the writing of simplified equations.

[134] *Castelluccio*, "small castle". The word obviously refers to the marginal calculations on the previous page, well enclosed by lines as if composed of several courtyards or chambers. Here, as we see, Benedetto states explicitly that the marginal calculation is already there when the main text is written.

[135] In some of Benedetto's preceding problem solutions with two algebraic unknowns, these are the *chosa*, "thing", and *quantità*, "quantity", cf. above. In others, his unknowns are *quantità* and *borsa*. As we see, Benedetto starts from the first routine and then chooses the other, allowing him to conserve the *borsa*.

quantities and $^{19}\!/_{25}$ of quantity.[136] Now it is to be seen if the *first* and *second* with the *purse* have two times as much as the *third*. And you shall say, the *first* has 4 *quantities* $^{827}\!/_{1100}$ less 1 *purse* $^{19}\!/_{25}$. And the *second* has 1 *quantity*. These two amounts, having been joined with the *denari* of the *purse*, make 5 *quantities* $^{827}\!/_{1100}$ less $^{19}\!/_{25}$ of *purse*. And this is two times as much as the *denari* of the *second*,[137] that is, 2 times as much as $^{13}\!/_{20}$ *quantity* and $^4\!/_5$ of *purse*, which are $^{26}\!/_{20}$ of *quantity* and $^8\!/_5$ of *purse*. So 5 *quantities* $^{827}\!/_{1100}$ less $^{19}\!/_{25}$ of *purse* are equal to $^{26}\!/_{20}$ of *quantity* $^8\!/_5$ of *purse*, where you shall confront (*raguaglerai*[138]) the sides detracting on both sides $^{26}\!/_{20}$ of *quantity* and giving to each side $^{19}\!/_{25}$ of *purse*. You shall have that 3 *quantities* $^{1597}\!/_{1100}$ are equal to $^{59}\!/_{25}$ of *purse*. And in order not to have fractions, multiply each side by 1100. You shall have that 2897 *quantities* are equal to 2596 *purses*. Therefore, when the *purse* is worth 4897, the *quantity* is worth 2596. And for the *first*, who has 4 *quantities* $^{823}\!/_{1100}$ less a *purse* $^{19}\!/_{25}$, he will have 3717. And the *second*, whom we posited to have a *quantity*, will have 2596. And the *third*, whom we found to have $^{13}\!/_{20}$ *quantity* $^4\!/_5$ *purse*, will have 5605. And the fourth, whom we found to have $^{273}\!/_{1100}$ of *quantity* $^4\!/_5$ of *purse*, had 4366. And thus it has been made, The *first* has 3717. And the *second* has 2596. And the *third* has 5605. And the fourth 4366. And the *purse* had 4897. Which further, [(fol. 271v)]reduced to smaller numbers by 59: The *first* has 63, the *second* 44, the *third* 95, the fourth man 74. And the *purse* 83.

Close attention to the organization of fol. 270[v] (redrawn in Figure 7) shows that the statement "Four have *denari* [...] how many *denari* there were in the purse" was written first. After that Benedetto started calculating in a "margin" which in certain points invades the text column by more than 80%. When that was done, he wrote a description of the calculations in whatever remained for that purpose.

The marginal calculation uses standard abbreviations for "first", "second", "third" and "fourth", which for typographical convenience we may render α, β, γ and δ. *Borsa*, *purse*, is abbreviated *b*.

The calculation starts in the upper left corner. After that, however, it is no longer linear. The written text assists in clarifying the order in which the different parts are to be read; Benedetto himself of course knew the order he was following.

At first (upper left corner of the "castelet") we find these four equations,[139]

(1) $\gamma = \frac{1}{2}\alpha + \frac{1}{2}\beta + \frac{1}{2}b$,

(2) $\delta = \frac{1}{3}\beta + \frac{1}{3}\gamma + \frac{1}{3}b$,

(3) $\alpha = \frac{1}{4}\gamma + \frac{1}{4}\delta + \frac{1}{4}b$,

(4) $\beta = \frac{1}{5}\alpha + \frac{1}{5}\delta + \frac{1}{5}b$,

---

[136] Error for "$^{19}\!/_{25}$ of purse". The marginal calculation, correctly, gives "δ $^{272}\!/_{1100}q$ $^{19}\!/_{25}b$".

[137] Error for "the third", as confirmed by the following words.

[138] As we see, the process of "confronting"/*ragugliamento* refers to the process of constructing the *reduced* equation – in the present case by addition as well as subtraction.

[139] Equality is indicated by large distance, addition by close juxtaposition. Elsewhere but not here Benedetto uses the long stroke.

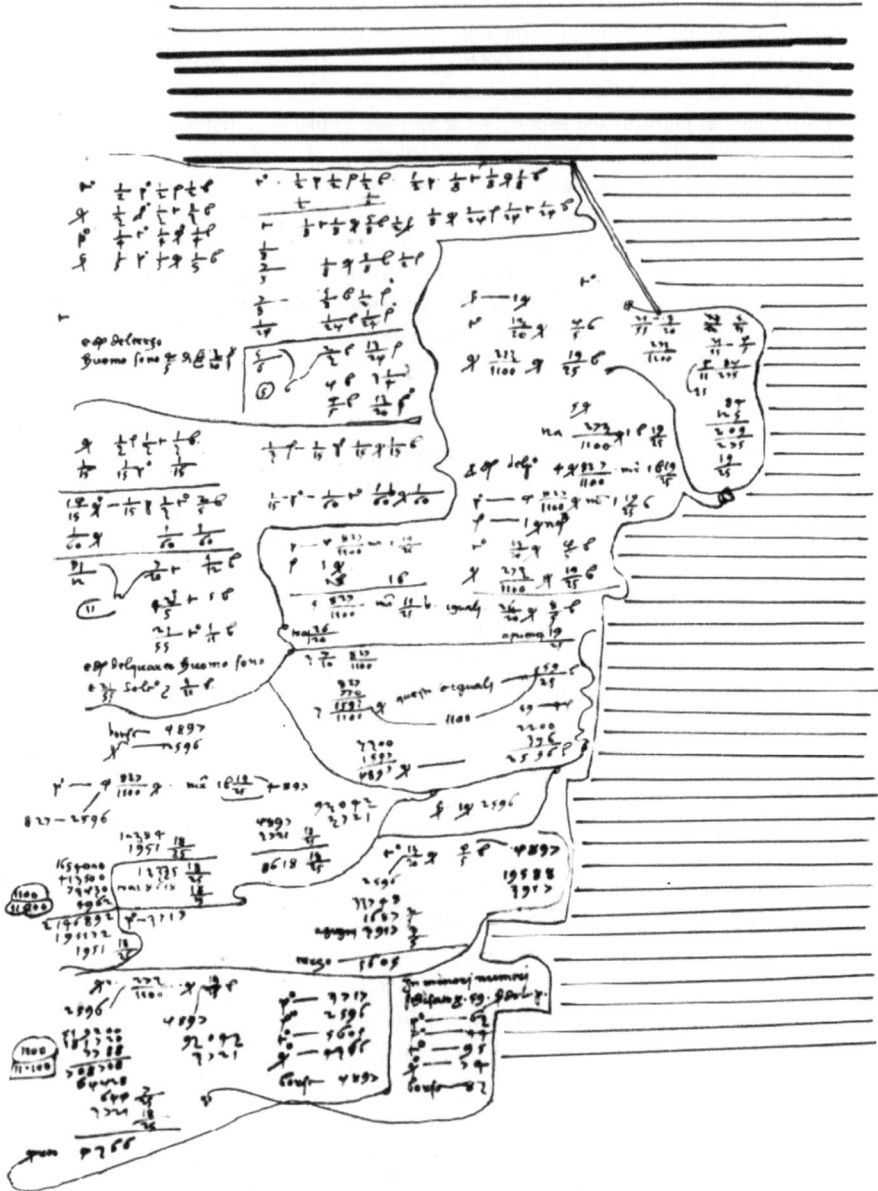

Figure 7. Fol. 270ᵛ, redrawn. Thick lines represent the problem statement, thin lines the procedure description (the first two lines belong to the procedure of the previous problem).

derived from the initial conditions of the problem via multiplication.

As a first step, in the same chamber, $\frac{1}{2}\alpha$ is found from (3) and substituted in (1), yielding

(5) $\gamma = \frac{1}{8}\gamma + \frac{1}{8}\delta + \frac{1}{8}b + \frac{1}{2}\beta + \frac{1}{2}b$.

This is reduced to

(6) $\frac{7}{8}\gamma = \frac{1}{8}\delta + \frac{1}{2}\beta + \frac{5}{8}b$.

Next (2) is used to find $\frac{1}{8}\delta$, which is substituted into (6), leading to

(7) $\frac{7}{8}\gamma = \frac{1}{24}\beta + \frac{1}{24}\gamma + \frac{1}{24}b + \frac{1}{2}\beta + \frac{5}{8}b$.

This is reduced to

(8) $\frac{5}{6}\gamma = \frac{13}{24}\beta + \frac{2}{3}b$,

which through division by $\frac{5}{6}$ is transformed into

(9) $\gamma = \frac{13}{20}\beta + \frac{4}{5}b$.

In the next chamber downwards, $\frac{1}{3}\beta$ is found from (4) and inserted into (2), which leads to

(10) $\delta = \frac{1}{15}\alpha + \frac{1}{15}\delta + \frac{1}{15}b + \frac{1}{3}\gamma + \frac{1}{3}b$.

This is reduced to

(11) $\frac{14}{15}\delta = \frac{1}{15}\alpha + \frac{1}{3}\gamma + \frac{2}{5}b$.

Now (3) is used to derive $\frac{1}{15}\alpha$, which is substituted into (11). That gives

(12) $\frac{14}{15}\delta = \frac{1}{60}\gamma + \frac{1}{60}\delta + \frac{1}{60}b + \frac{1}{3}\gamma + \frac{2}{5}b$,

which reduces to

(13) $\frac{11}{12}\delta = \frac{7}{20}\gamma + \frac{5}{12}b$.

Division by $\frac{11}{12}$ reduces this to

(14) $\delta = \frac{21}{55}\gamma + \frac{5}{11}b$.

Similarly to what was done in the egg-orange problem in the Ottoboniano *Praticha*, the large chamber to the right now shifts to standard *modo recto*, the *quantità* or *q* being identified with β, and *borsa* or *b* remaining in service. From (9) we get that

(15) $\gamma = \frac{13}{20}\beta + \frac{4}{5}b$,

and from (14) that

$$\delta = \frac{21}{55}\cdot(\frac{13}{20}q + \frac{4}{5}b) + \frac{5}{11}b ,$$

whence

(16) $\delta = \frac{273}{1100}q + \frac{19}{25}b$.

Further, from the original condition behind (4) we know that

$$\delta + \alpha + b = 5\cdot\beta ,$$

whence

(17) $\delta + \alpha + b = 5q$.

Therefore

$$\alpha = 5q - (\frac{273}{1100}q + 1\frac{19}{25}b) ,$$

or

(18) $\alpha = 4\frac{827}{100}q - 1\frac{19}{25}b$.

Inserting values for α and β in the original condition that gave rise to (1),

(19) $2\gamma = \alpha + \beta + b$,

we get

(20) $$2\gamma = 5\,^{827}/_{1100}\,q - {}^{19}/_{25}\,b\ ,$$

that is,

(21) $$5\,^{827}/_{1100}\,q - {}^{19}/_{25}\,b = {}^{26}/_{20}\,\beta + {}^{8}/_{5}\,b$$

Addition and subtraction lead to

(22) $$3\,^{1597}/_{1100}\,q = {}^{59}/_{25}\,b\ .$$

Multiplying by 1100 "so as to avoid fractions" Benedetto gets

(23) $$4897q = 2596b\ .$$

So (lower left chamber), if b is chosen to be 4897 (as Benedetto knows, the problem is indeterminate and allows this choice), q will be 2596. From (18) then follows that $\alpha = 3717$; β is already known to be 2596; γ can be found for instance from (15) to be 5605, and δ (16) to be 4366; b is already known to be 4897.

Since the problem is indeterminate and Benedetto's value for b was a *choice*, all values can be changed proportionally. Therefore (lower right chamber) Benedetto reduces them by the common factor 59, which gives him $\alpha = 63$, $\beta = 44$, $\gamma = 95$, $\delta = 74$, $b = 83$.

Comparison of the marginal calculation with the text that explains the procedure shows that the latter contains a number of typical copying errors, confirming that the marginal calculation was made first. The marginal calculation thus presents us with a clear instance of *incipient symbolic algebra* involving five unknowns, antedating other known examples by a small century.

Benedetto does not present what he has done as epoch-making, even though the purpose for which he is writing would invite that – in a gift intended for a protector (above, note 51), it would be obvious to show the merit of what is offered indirectly by pointing out with due modesty that something never made before is offered.

As we shall see, Benedetto *was* aware to have produced an innovation. A comparison with Fibonacci's *Liber abbaci* may tell us why he may have considered this innovation marginal. In chapter 12 part 4 of Fibonacci's work [ed. Giusti 2020: 372] we find this problem:

> On four men and a purse.
> The first and the second with the purse have the double of the *denarii* of the third; and the second and the third the triple of the fourth, and then the third and the fourth the quadruple of the first, while the fourth and the first with the purse similarly have the quintuple of the second.

This is obviously the same problem, and the rest of Benedetto's *Praticha* leaves no doubt that Benedetto knew the *Liber abbaci*. Fibonacci goes on:

> The solution to this problem you will find by finding the ratio of the *denarii* of the purse to the *denarii* of the first in this way. Because the first and second with the purse have the double of the third, half of the *denarii* of the first and second and the purse is as much as the *denarii* of the third man. Similarly from the other propositions you will have that $^{1}/_{3}$ of the second and third man and of the purse is as much as the *denarii* of the fourth

man, and $\frac{1}{4}$ of the third and fourth man and of the purse is the quantity of the *denarii* of the first, and $\frac{1}{5}$ of the *denarii* of the fourth and first man and of the purse is the quantity of the *denarii* of the second. And because $\frac{1}{2}$ of the first and second and of the purse is the quantity of the third, the third part of the first and second and purse, that is $\frac{1}{6}$ of them, is $\frac{1}{3}$ of the third man. Commonly (*comuniter*) are joined $\frac{1}{3}$ of the *denarii* of the second and purse: then will $\frac{1}{6}$ of the first and $\frac{1}{2}$ of the second and of the purse be as much as $\frac{1}{3}$ of the second and third and of the purse. But $\frac{1}{3}$ of the second and third and of the purse is the quantity of the *denarii* of the fourth man; hence $\frac{1}{6}$ of the first and $\frac{1}{2}$ of the second and of the purse are the quantity of the *denarii* of the fourth man. Therefore $\frac{1}{4}$ of $\frac{1}{6}$ of the *denarii* of the first, that is, $\frac{1}{24}$, and $\frac{1}{4}$ of $\frac{1}{2}$, thus $\frac{1}{8}$ of the *denarii* of the second and of the purse, are $\frac{1}{4}$ of the *denarii* of the fourth man. Commonly are added $\frac{1}{4}$ of the third and of the purse: then $\frac{1}{24}$ of the first with $\frac{1}{8}$ of the second and with $\frac{1}{4}$ of the third and $\frac{3}{8}$ of the purse will be as much as $\frac{1}{4}$ of the *denarii* of the third and fourth and of the purse. But $\frac{1}{4}$ of the third man and the fourth and of the purse is the quantity of the first. Therefore $\frac{1}{24}$ of the first and $\frac{1}{8}$ of the second and $\frac{1}{4}$ of the third and $\frac{3}{8}$ of the purse are as much as the *denarii* of the first. Then their fifth part, that is $\frac{1}{120}$ of the first and $\frac{1}{40}$ of the second and $\frac{1}{20}$ of the third and $\frac{3}{40}$ purse, are $\frac{1}{5}$ of the *denarii* of the first. Commonly are added $\frac{1}{5}$ of the fourth man and the purse: then $\frac{1}{120}$ of the first and $\frac{1}{40}$ of the second and $\frac{1}{20}$ of the third and $\frac{1}{5}$ of the fourth and $\frac{11}{40}$ of the purse will be as much as $\frac{1}{5}$ of the fourth man and the first and of the purse. [...]

The omission [...] is as long as the part that was translated. It leads to

Hence $\frac{79}{600}$ and $\frac{1}{150}$ of the first, that is $\frac{83}{600}$ of the same, with $\frac{1}{25}$ of the purse, are $\frac{29}{200}$ of the purse. Commonly are taken away $\frac{1}{25}$ of the purse. Remain $\frac{83}{600}$ of the first, as much as $\frac{21}{200}$ of the purse. Then two numbers should be found so that $\frac{83}{600}$ of the first are $\frac{21}{200}$ of the second, they will be 63 and 83. Then if the first man has 63, the purse is 83. [...].

If we admit the identity of "the *denari* of the first/first man", "the quantity of the *denari* of the first man", "the quantity of the first man" and "the first man", this is rhetorical algebra with five unknowns. If we insist on fully consistent and not only unmistakeable naming it may not be, but the difference is scant, and would hardly have been thought of at the time.[140]

There are traces in Fibonacci's text showing that even he *described* a procedure performed by other means: "150" instead of "$\frac{1}{150}$ *primi*" and "*denariis secondi*" instead of "*denariis primi*". Both are shared by all manuscripts, showing that they were in Fibonacci's master copy – see [Høyrup 2021b: 5]. This may have been a line diagram similar to the one used in the first solution to the give-and-take problem whose alternative solution served to introduced the *regula recta* – in all cases where these diagrams are shown

---

[140] The same objection could be raised to Benedetto's verbal description of the procedure. Since this description is secondary it changes nothing in the characterization *of the marginal calculation* as a perfect symbolic algebraic calculation with five unknowns.

in his text they are lettered *a-b-g-d*, meaning that they belong with a faithfully borrowed text (see [Høyrup 2021b: 8–11]). We may suppose that he used the same technique when working independently, but in those cases he hid it.

Fibonacci solves many intricate horse-, give-and-take- and purse-finding problems with similar forbiddingly difficult rhetorical arguments. We may guess that all of these were solved by similar means, which it would be difficult not to characterize as algebraic (there should be no essential difference between representing unknown numbers by line segments and by letters). If this is true, Fibonacci was deeply engaged in linear algebra with many unknowns, but only in secret.

Benedetto in any case did not know; but he will have been aware that the difference between what he had done himself and what could be found in the *Liber abbaci* was scant. He will therefore have had sound reasons to see his innovation as minor.

Benedetto's awareness of having none the less innovated is shown by his treatment of two horse-buying problems found later in the *Praticha*.

On fol. 277ʳ we find this:[141]

(fol. 277r)Four men have *denari* and want to buy a horse, and no one has so many *denari* that he can buy it. The first says to the second and the third, if you give me ½ of your *denari*, with mine I shall buy the horse. The second says to the third and fourth man, if you give me the ⅓ of your *denari*, with mine I shall buy the horse. The third man says to the fourth and the first, if you give me the ¼ of your *denari*, I shall buy the horse. Further, the fourth man asks the first and the second for the ⅕ of their *denari* and says to buy the horse. It is asked, how many *denari* each one had, and what the horse was worth.

Even though there are many ways to solve such cases I shall take the most convenient, or let us say the least tedious.‖ That is that you shall say, we propose that the first with the half of the *denari* of the second and of the third man has a horse. And we say that the second with the third of the *denari* of the third and fourth man buy the horse. So the first with the ½ of the *denari* of the second and third man has as much as the second has with ⅓ of the third and fourth man. From there you will see confronting (*raguagliando*), that is, detracting first on each side (*parte*) ½ of the *denari* of the second, and we shall have that the *denari* of the first with ½ of the *denari* of the third man are as much as ½ of the *denari* of the second with ⅓ of the third and fourth man. And then you remove from each side ⅓ of the *denari* of the third man, and we shall have that the first man with ⅙ of the *denari* of the third man is as much as ½ of the *denari* of the second. And take note of this (*nota*). [...]

Until the point marked ‖, the text was written first. Then Benedetto starts calculating in the margin, at first in these steps (using the same standard abbreviations as before for "first", "second", "third" and "fourth"):[142]

---

[141] A closely related problem is found in the *Liber abbaci* [ed. Giusti 2020: 393]. There too, Fibonacci's procedure is at least quasi-algebraic, but once again Benedetto calculates on his own.

[142] Addition is still indicated by juxtaposition, but now equality first by "*iguali*by ——————— alone",

$$\alpha + \tfrac{1}{2}\beta + \tfrac{1}{2}\gamma = \beta + \tfrac{1}{3}\gamma + \tfrac{1}{3}\delta$$
$$\alpha + \tfrac{1}{2}\gamma = \tfrac{1}{2}\beta + \tfrac{1}{3}\gamma + \tfrac{1}{3}\delta$$
$$\alpha + \tfrac{1}{6}\gamma = \tfrac{1}{2}\beta + \tfrac{1}{3}\delta$$

$$\beta + \tfrac{1}{3}\gamma + \tfrac{1}{3}\delta = \gamma + \tfrac{1}{4}\delta + \tfrac{1}{4}\alpha$$
$$\beta + \tfrac{1}{3}\delta = \tfrac{2}{3}\gamma + \tfrac{1}{4}\delta + \tfrac{1}{4}\alpha$$
$$\beta + \tfrac{1}{12} = \tfrac{2}{3}\gamma + \tfrac{1}{4}\alpha$$

The structure of the marginal calculation is similar to that of the previous example – divided into sections, the first of these (redrawn here) written close to the margin and not occupying much of the text column, those written later then spreading further into it. There is no need to say more about this.

The calculation was thus made first even this time, and the describing text written afterwards. In the next passage we see Benedetto sharpening of his conceptual apparatus, speaking explicitly about the (reduced) equations and giving a name to the isolation of one unknown:

> And then you shall say, we have said that the second man with $\tfrac{1}{3}$ of the *denari* of the third and fourth man has as much as the third man with $\tfrac{1}{4}$ of the *denari* of the fourth and first man. Where the *denari* of the second man with $\tfrac{1}{3}$ of the *denari* of the third and fourth man are as many as are the *denari* of the third man with $\tfrac{1}{4}$ of the *denari* of the fourth and first (fol. 277v) man. Where confronting the sides you take away $\tfrac{1}{3}$ of the *denari* of the third man, and you shall have the *denari* of the second with the $\tfrac{1}{3}$ of the *denari* of the fourth to be the $\tfrac{2}{3}$ of the *denari* of the third man with $\tfrac{1}{4}$ of the *denari* of the fourth and first man. And then on each side you take away $\tfrac{1}{4}$ of the fourth man. And you shall have that the *denari* of the second man with $\tfrac{1}{12}$ of the fourth man are as much as $\tfrac{2}{3}$ of the third man with $\tfrac{1}{4}$ of the *denari* of the first man. Which you still take note of. And in this way you may

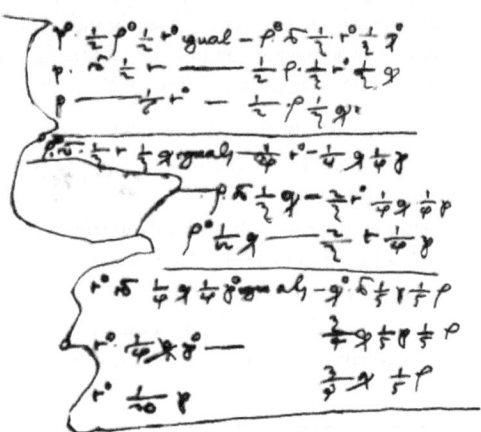

Figure 8. The first part of the calculations on fol. 277ʳ, redrawn.

then by ——————— alone.

confront[143] the other positions. But with these 2 you can solve. And if you want the other equations,[144] you may do as you see on the previous page, there the equation of the third is.[145] Now to our subject-matter (*materia*). We have made that the *denari* of the first with $\frac{1}{6}$ of the *denari* of the third man are $\frac{1}{2}$ of the *denari* of the second and $\frac{1}{3}$ of the *denari* of the fourth man. And we also have that the second with $\frac{1}{12}$ of the fourth man are as much as the $\frac{2}{3}$ of the *denari* of the third man and $\frac{1}{4}$ of the *denari* of the first. Therefore $\frac{1}{4}$ of the *denari* of the first is to be brought apart from the *denari* of the others. You shall keep this way, we have that the *denari* of the first and $\frac{1}{6}$ of the *denari* of the third man are as much as the $\frac{1}{2}$ of the *denari* of the second and $\frac{1}{3}$ of the *denari* of the fourth man, where from both sides you take away $\frac{1}{6}$ of the *denari* of the third man. And you shall have the *denari* of the first to be $\frac{1}{2}$ of the *denari* of the second and $\frac{1}{3}$ of the *denari* of the fourth less $\frac{1}{6}$ of the *denari* of the third man. Therefore $\frac{1}{4}$ of the *denari* of the first man are $\frac{1}{8}$ of the *denari* of the second and $\frac{1}{12}$ of the *denari* of the *denari* of the fourth and less $\frac{1}{24}$ of the *denari* of the third man. And that you shall join to $\frac{2}{3}$ of the *denari* of the third man, and you shall have $\frac{5}{8}$ of the *denari* of the third man and $\frac{1}{8}$ of the second and $\frac{1}{12}$ of the fourth man. Therefore you shall say that the *denari* of the second with $\frac{1}{12}$ of the *denari* of the fourth are as much as $\frac{5}{8}$ of the *denari* of the third man and $\frac{1}{12}$ of the *denari* of the fourth man and $\frac{1}{8}$ of the *denari* of the second. Therefore, from each side you shall take away $\frac{1}{8}$ of the second man and $\frac{1}{12}$ of the fourth man. You shall have that $\frac{5}{8}$ of the *denari* of the second are as much as $\frac{5}{8}$ of the *denari* of the third man. Now this is known, you shall say, if the second man should have 5, then the third would have 7. And having had this insight (*lume*), and we shall make position that the second man had 5 quantities, it follows that the third man would have 7 quantities. [...].

From here onward, in the margin and as described in the text, Benedetto makes (symbolic respectively rhetorical) algebra with two unknowns only, and we do not need to follow him.

After a simpler problems solved rhetorically by means of the unknowns *quantità* and *chavallo* then comes (fol. 278ᵛ) another problem taken over from Fibonacci [ed. Giusti 2020: 397]. Here Benedetto brings his new method into play. He *might* have expressed Fibonacci's quasi-algebraic procedure by means of his new technique, and does so at first – not necessarily copying, these are the obvious first steps. Then, however, the two solutions

---

[143] Or "make (reduced) equations of" – I have been unable to find an English translation reflecting both senses.

[144] *Aguagliazioni* – that is, the (reduced) equations resulting from the process of *ragugliamento*, "confrontation".

[145] Namely in the three lines of the marginal calculation that follow immediately after the two times three lines rendered above – that is, the last three lines of the manuscript excerpt in Figure 8. We may presume that Benedetto after having made all three sets found out in the ensuing calculations that he did not need the last of them. We may remember his words "It is true that it could be done without that, but since it is in the castelet, this order shall be pursued" in his solution to the problem discussed previously.

diverge.

Four have *denari* for which they want to buy a
horse, and none of them has so many *denari* that
he can buy it. The first and the second say to the
third man, if you give us the $\frac{1}{3}$ of your *denari*,
we shall buy the horse. The second and third man
say to the fourth man, if you give us the $\frac{1}{4}$ of
your *denari* we shall buy the horse. The third and
fourth man say to the first, if you give us the
$\frac{1}{5}$ of your *denari*, with ours we shall buy the
horse. The third and first man say to the second,
if you give us the $\frac{1}{6}$ of your *denari* we shall buy
the horse. It is asked, how much each one had,
and what the horse was worth. We shall do it by
equation.[146] Where you shall say, the first and
second with $\frac{1}{3}$ of the third buy the horse. And
the second and third man with $\frac{1}{4}$ of the fourth
man buy the horse. Thus the *denari* of the first
and second with $\frac{1}{3}$ of the *denari* of the third man
are as much as are the *denari* of the second and
third man with $\frac{1}{4}$ of the *denari* of the fourth.
Where confronting the sides, taking away from
each side the *denari* of the second and $\frac{1}{3}$ of the
denari of the third, we shall have that the *denari*
of the first are as much as $\frac{2}{3}$ of the *denari* of the
third man and $\frac{1}{4}$ of the *denari* of the fourth man.
And mark this. Then you shall say, the second
and third man with $\frac{1}{4}$ of the *denari* of the fourth
man buy the horse. And the third and fourth man
with $\frac{1}{5}$ of the *denari* of the first buy a horse. So
the *denari* of the second and third man with
$\frac{1}{4}$ of the *denari* of the fourth man are as much
as the *denari* of the third and fourth man with
$\frac{1}{5}$ of the *denari* of the first. Therefore take away
from each side the *denari* of the third and $\frac{1}{4}$ of
the *denari* of the fourth man, and we shall have
that the *denari* of the second are $\frac{3}{4}$ of the *denari*
of the fourth and $\frac{1}{5}$ of the *denari* of the first.
And then, going on, you shall say that the third
and fourth man with $\frac{1}{5}$ of the *denari* of the first
buy the horse. And the fourth and first with $\frac{1}{6}$

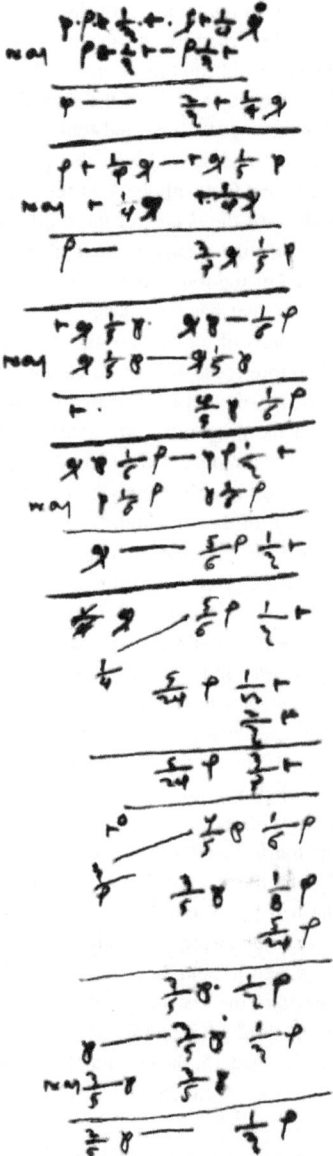

Figure 9. Benedetto's marginal calculation on fol. 278^v, redrawn.

---

[146] *per aguagliatione.*

of the *denari* of the second buy the horse. It therefore follows that the *denari* of the third and fourth man with $\frac{1}{5}$ of the *denari* of the first are as much as the first and fourth with $\frac{1}{6}$ of the *denari* of the second. Where, confronting the sides, taking away on each side the *denari* of the fourth man and $\frac{1}{5}$ of the first, we shall have that the third man has the $\frac{4}{5}$ of the first and $\frac{1}{6}$ of the second. And mark this. And thus you shall do for the fourth man, saying, the first and fourth with the $\frac{1}{6}$ of the *denari* of the second buys the horse. The first and second with the $\frac{1}{3}$ of the *denari* of the third buy the horse. Therefore the fourth and first with the $\frac{1}{6}$ of the *denari* of the second have as much as the first and second with $\frac{1}{3}$ of the *denari* of the third man. Therefore confronting the sides, taking away on each side the *denari* of the first and $\frac{1}{6}$ of the *denari* of the second, we shall have that the *denari* of the fourth are $\frac{5}{6}$ of the *denari* of the second and $\frac{1}{3}$ of the *denari* of the third. And of that has been taken note. And you shall begin at the first equation,[147] saying, the *denari* of the first are the $\frac{2}{3}$ of the *denari* of the third man and $\frac{1}{4}$ of the fourth man. Therefore it has to be known what $\frac{1}{4}$ of the *denari* of the fourth are. From the others, however, we have found that the *denari* of the fourth man are the $\frac{5}{6}$ of the second and $\frac{1}{3}$ of the third man, where the $\frac{1}{4}$ of the *denari* of the fourth man are as much as the $\frac{5}{24}$ of the *denari* of the second and $\frac{1}{12}$ of the *denari* of the third. Where to the $\frac{2}{3}$ of the *denari* of the third man you join $\frac{1}{12}$ of the *denari* of the third and $\frac{5}{24}$ of the second, they make $\frac{3}{4}$ of the third and $\frac{5}{24}$ of the *denari* of the second. And then bring the $\frac{3}{4}$ of the third apart from the others, saying, the third man has the $\frac{4}{5}$ of the first and $\frac{1}{6}$ of the second, where the $\frac{3}{4}$ of the third man are the $\frac{3}{5}$ of the first and $\frac{3}{24}$ of the second. And you shall join to $\frac{5}{24}$ of the second $\frac{3}{5}$ of the first and $\frac{3}{24}$ of the second, they make $\frac{3}{5}$ of the first and $\frac{1}{3}$ of the second, and we shall have made that the *denari* of the first are as much as $\frac{3}{5}$ of the first and $\frac{1}{3}$ of the second. Therefore you shall detract on both sides the *denari* of the first, you shall have that $\frac{2}{5}$ of the *denari* of the first are $\frac{1}{3}$ of the *denari* of the second. That is, that the $\frac{2}{5}$ of the *denari* of the first are as much as the $\frac{1}{3}$ of the *denari* of the second. Thus, if the first should have 5, the second would have 6. Let us now try the others. You shall say that the third has as much as the $\frac{4}{5}$ of the first and the $\frac{1}{6}$ of the second. Therefore, the $\frac{4}{5}$ of the first and $\frac{1}{6}$ of the second are 5. And so much would he have. And the fourth has $\frac{5}{6}$ of the second and $\frac{1}{3}$ of the third, where the $\frac{5}{6}$ of the second are 5 and $\frac{1}{3}$, and the $\frac{1}{3}$ of the third man are $1\frac{2}{3}$, [fol. 279r]which all make $6\frac{2}{3}$. And thus it is done, the first has 5 and the second 6 and the third 5 and the fourth $6\frac{2}{3}$. Which, so as not to have fractions, multiply all by 3. And you shall have the first 15, the second 18, and the third man 15, and the fourth man 20. And so as to know what the horse is worth, you shall join 15 of the first and 18 of the second, they make 33. To these joined the $\frac{1}{3}$ of the *denari* of the third man, that is, of 15, they make 38. And as much is worth the horse. And thus the first had 15, the second 18, and the third had 15, and the fourth man had 20. And the horse was worth 38.

Now the technique is mature. This time Benedetto resists the temptation to shift to the traditional two unknowns, he uses the four unknowns (the price of the horse does not

---

[147] *aguagliatione* – the first *reduced* equation.

enter in the algebraic manipulations) until the very end. Moreover, the marginal calculation is extremely neat – see Figure 9. It fills only a narrow column in the margin and does not go into the text column (corrected errors in the describing text show that the marginal calculations were still made first).

A related problem follows on fol. 279^r. Even this, Benedetto says, will be made "by equation", and he shows the construction of the first reduced equation. For the others he refers to "the teaching made below", namely in a space of 11.5 × 9 centimetres. Unfortunately it has not been filled by calculations, but it is none the less clear that Benedetto thought that his symbolic calculations would be preferable to a verbal description.

It is possible that a few problems presented without detailed calculations were meant to be dealt with by the new method. That, however, was the end of it. There is no use for it in the rest of the *Praticha*. The incomplete copies of the *Praticha* that have survived do not contain it, and even Ghaligai who knows Benedetto's treatise has not noticed (even his manuscript may have been incomplete). All in all, a glorious failure.

Pacioli, Chuquet and de la Roche

So it seems at least, judging from known surviving sources. As Pacioli reveals, however, known surviving sources do not give a complete picture. It is not totally to be excluded that somebody at the time was inspired by Benedetto's invention. In the long run, however, it left no traces. For a while, nobody would work with more than two unknowns, nor make complete symbolic calculations.

Two "horse"-problems in Pacioli's Perugia manuscript [ed. Calzoni & Gavazzoni 1996: 311–312] make use of the algebraic unknowns *thing* and *horse*. They explain that *horse* is nothing but the price of the horse, and posit in both problems that the first man has a *thing* and the other two together 2 *horses* less 2 *things* (given that in both problems the first, having received half of what they have together, will have 1 *horse*).

This seems similar to what Fibonacci and Benedetto had done in similar problems. The rest of the calculation, however, does not coincide with what we have seen so far: in order to discover the ratio between the possessions of the second and the third, twice a new *thing* is introduced (with no distinction of name).

That trick is used again and better explained in the *Summa*, to which we shall turn. Here, the use of a second unknown called *quantità* (abbreviated q^a) is dealt with twice as a minor topic.

On fol. 148^v, the *cosa* as well as the *quantità* are spoken of as *doi quantità sorde*, "two deaf quantities". They are put into play in a numerical variant of Antonio's #9 (above, p. 80) – here

$$AB = 8 \ , \ A^2 + B^2 = 20 \ .$$

Since #9 is the problem where Antonio first approaches the use of two unknowns with hesitation while Pacioli makes use of the fully developed method, the borrowing is almost certainly indirect; though we possess no intermediate sources, they must have existed.

Returning to the topic in earnest on fol. 191^{v} Pacioli says he explains it "only to show how one operates with a deaf quantity which the ancients called second things so as to distinguish them from the first position". He then actually operates with *three* unknowns, eliminating however the second before introducing the third; this allows him to recycle the name *quantità* (I use Pacioli's notation, where *co* stands for the *thing*):

> Three have *denari*. The first says to the other 2, if you give me half of yours I shall have 90. The second says to the other 2, if you give me $\frac{1}{3}$ of yours, I shall have 84. The third says to the other 2, if you give me $\frac{1}{4}$ of yours plus 6, I shall have 87. I give you this solely to show how one operates by means of a deaf quantity which the ancients call second *thing* to differentiate it from the first positions. Posit that the first has 1 *co*, remove it from 90, 90 less 1 *co* remains, and this should be $\frac{1}{2}$ of the other 2, these then have 180 less 2 *things*, and all 3 have 180 less 1 *co*. Now do for the 2nd and posit that he has a *quantity*, which I depict thus, one $\tilde{\Phi}^{s}$, and to the 2 remain 180 less a *co* less a $\tilde{\Phi}^{s}$. Take $\frac{1}{3}$, from which results 60 less $\frac{1}{3}co$ less $\frac{1}{3}\tilde{\Phi}^{s}$.

If *A*, *B* and *C* stand for the three possessions, the conditions are thus

$$A+\tfrac{1}{2}(B+C) = 90 \; , \quad B+\tfrac{1}{3}(A+C) = 84 \; , \quad C+\tfrac{1}{4}(A+B)+6 = 87 \; .$$

With *A* posited to be a *co*, *B* to be a $\tilde{\Phi}^{s}$, Pacioli has thereby found that

$$\tfrac{1}{3}(A+C) = 60-\tfrac{1}{3}co - \tfrac{1}{3}\tilde{\Phi}^{s} \; .$$

Inserting this in $B+\frac{1}{3}(A+C) = 84$ and using that $B = 1\tilde{\Phi}^{s}$, Pacioli derives the equation

$$1\tilde{\Phi}^{s} = 36+\tfrac{1}{2}co \; ,$$

which is the second possession.

Now comes something new:

> Now for the 3d do similarly: Posit that he has a *quantity*, remove it from 180 less 1 *co*, that is, still from the amount of all three. [...].

Pacioli thus operates with three algebraic unknowns, but only with two at a time, which allows him to recycle the name *quantity*. This second position allows Pacioli to derive the equation

$$1\tilde{\Phi}^{s} = 48+\tfrac{1}{3}co \; ,$$

which is the third possession. That brings him back to a single unknown,

$$A+B+C = 1co+36+\tfrac{1}{2}co+48+\tfrac{1}{3}co \; .$$

But we know that $A+B+C = 180-1co$. This solves the problem. In the end Pacioli specifies that one shall always with this method isolate the *quantity*, and explains that

> by means of these deaf quantities which the ancients called second *things* a great many strong problems can be solved by the one who handles the equations well.

Nobody knows the identity of Pacioli's "ancients", but Pacioli's reference to them shows that two unknowns had been practised in abbacus algebra more broadly than we might believe on the basis of surviving sources – none of those which we have looked at and which employ a second or further unknowns ever give a name to the category.

That they must have existed is confirmed by Chuquet – see [Heeffer 2012: 134*f*]. He uses the same method of a recycled second unknown repeatedly in the appendix to his *Triparty* (written in Lyons in 1484, as we remember from p. 35). Since the *Triparty* was a manuscript and apparently did not circulate (apart from de la Roche's use of kind of fair copy in [1520]), Chuquet was certainly not Pacioli's source. On the other hand, the shared principle supports Pacioli's claim that he presents existing ideas. This is one of the things de la Roche takes over from Chuquet, under the heading *règle de la quantité* [1520: 61^r].

De la Roche says that this rule

> is inserted in the first canon of the rule of the thing as accomplishment and perfection of the same because it often happens in several problems from this canon that one must posit two or more times, of which the first position is 1 ρ. If it happens afterwards that one must posit another time or two or three, etc., it is needed that the second, third or fourth positions are different from the number ρ.

Giving the same name to all would indeed lead to confusion.

The first example offered by de la Roche is strange:

> Divide 1 ½ ρ plus 15 in two parts such that if one removes 13 from the second in order to add it to the first, this first part will be the quadruple of the second plus 2.

It must be a borrowing from a complete problem dealing with a *thing* and a *quantity* (perhaps an imagined complete problem), but only of that part of the calculation where the second unknown is eliminated.

The following problems are more regular. They are mostly complicated version of horse-buying-, give-and-take- or similar problems in abstracted form; on fol. 122^r, there is a problem about the purchase of saffron, cinnamon and cloves, and on fol. 217^r there is a genuine horse-problem. Many use more than two unknowns but recycle the *quantité*. Taken together, they present only a small fraction of de la Roche's algebra, in spite of his praise of the importance of the *règle de la quantité*.

Heeffer supposes that all those who in subsequent years spoke of the *règle de la quantité* or *regula quantitatis* or who applied it were inspired by de la Roche. This is no certain conclusion. As pointed out by Heeffer [2012: 134], a marginal note in Chuquet's manuscript, apparently written by de la Roche (and in that case while he was preparing his own book) states that "ceste règle est appelée la Regle de la quantité", "this rule is called the Rule of the quantity". "... est appelée", not "I shall call" or "shall be called". De la Roche has thus recognized a method which he already knows under that name; there is no reason that everybody else who used the phrase should know it from de la Roche (e.g., Rudolff – but see imminently) – nor *a fortiori* that he should be the source for all those who worked with *quantitas* as a second unknown (e.g., Cardano, [1539] as well as [1545]).

Coß

The first to take up two unknowns within the *coß* tradition seems to be Rudolff. In spite of de la Roche's laudatory words, Rudolff is also likely to have been the first who really saw the operation with several unknowns not a peripheral possibility but as essential to algebra – in our perspective, first to *understand* it to be essential to *our* algebra.

Indeed, when introducing for the first time the *Regel quantitatis* on fol. I vi^r he asserts it to be "a completion of the *coss*, indeed in truth a completion without which it would not be worth much more than a trifle [*ein pfifferling*]". A second introduction comes on fol. P vi^v, before the next use of the rule; here he refers to the need to avoid "confusion" – close enough to de la Roche to make inspiration plausible.

The example that follows there changes plausibility into quasi-certainty. It asks for the splitting of $1\mathcal{R}+14\phi$ into two parts fulfilling a certain condition – that is, another subordinate part of a complete calculation (and, if inspired by de la Roche, definitely an imagined complete calculation since Rudolff cannot have known the full context).

There are many more problem solutions employing several unknowns than in de la Roche's book, a large part of which make use of a recycled *quantitet*. Of particular interest is one found on fol. P vii^v. It deals, *not* as always done in the tradition until now, with "two numbers" but with "two numbers *a* and *b*". *a* is then posited to be $1\mathcal{R}$, and *b* 1*quantitet*. Whereas the Ottononiano *Praticha* and Benedetto were fully able to perform algebraic operations on newly introduced symbols, that falls outside what Rudolff can accept within the *coß*. *Coß, his* algebra, operated with $\mathcal{R}$ and, when needed, with *quantitet*, and not with arbitrary symbols.

Stifel

That was to change, *though still within limits*, with Stifel's *Arithmetica integra* from [1544]. Stifel certainly knows Rudolff – anybody working on *coß* after 1525 did, and in [1553] Stifel was to republish an expanded edition of Stifel's *Coss*.

Stifel was the first algebra writer after Benedetto to create a system for naming more than two unknowns – and the first to *create a system*, namely from fol. 252^v onward in the *Arithmetica integra*. Stifel still thinks of $\mathcal{R}$, the *res*, as the primary unknown (his headline for the topic is *De secundis radicibus*, "on second roots"). For these second roots he uses the sequence of letters of the alphabet, "1A (that is, $1A\mathcal{R}$), 1B (that is, $1B\mathcal{R}$), 1C (that is, $1C\mathcal{R}$), 1D etc."; for their second powers he uses $1A\mathcal{z}$ etc. For the product of $\mathcal{R}$ and A he suggests $\mathcal{R}A$, while that of A and B will be written AB.[148] Stifel also

---

[148] The explanation of 1A as standing for $1A\mathcal{R}$ shows that A, B, etc. are thought of as markings. So, all the first powers are $\mathcal{R}$, but they are distinguished as $^A\mathcal{R}$ ("the A-kind of $\mathcal{R}$"), $^B\mathcal{R}$, etc. When standing to the left, on the other hand, $\mathcal{z}$ is meant as a factor. This system seems somewhat heavy and prone to produce mistakes; as we shall see, Stifel would soon give it up.

shows divisions of such products of powers of the unknowns (to use modern terms), yet only such that do not lead to negative powers.

A first example (fol. 252^v) is borrowed from Rudolff and makes use of only two unknowns. It is uninteresting on both accounts, involving nothing but a substitution of A for q (for this problem, admittedly, Rudolff employs only one unknown).

The next example (fol. 253^v) teaches us much more. It is mathematically simple, belonging to a type which we may speak of as "all except each":

> Seven men owe me money in this way. The first and second, third, fourth, fifth and sixth owe 142 florins. (Here observe, that only the debt of the seventh debtor is excluded from this amount of florins.) I posit therefore that the amount of the seventh is $1\varkappa$, and thus that the amount of all the debts will be $142+\varkappa$. The second, third, fourth, fifth, sixth and seventh owe 126 florins. (Here the debt of the first is excluded.) I posit therefore for the amount of the first $1A$ florins. And thus again the amount of all results, making $126+1A$. [...].

The problem formulation continues cyclically. The following numbers are therefore equal ($\varkappa$, A, B, C, D, E and F being Stifel's names for the respective debts):

$$142+1\varkappa$$
$$126+1A$$
$$136+1B$$
$$128+1C$$
$$130+1D$$
$$120+1E$$
$$148+1F$$

In Fibonacci's *Liber abbaci* as well as a in number of abbacus treatises, such problems are solved without recourse to algebra. In the present case their authors would have observed that the sum $142+126+136+...+148 = 930$ contains the sum of the debts seven times, less the same sum once. Dividing the sum by 6 therefore shows that the sum of the debts is 155.

But Stifel's primary interest is not to solve problems: he wants to illustrate a technique, and therefore the possibility to eschew algebra does not interest him. Nor is there any reason he should point out that the problem allows easy recycling of the secondary unknown. He therefore proceeds as follows:

From the equality of the first two amounts follows $A = 16+1\varkappa$. Similarly, $B = 6+1\varkappa$, $C = 14+1\varkappa$, $D = 12+1\varkappa$, $E = 22+1\varkappa$, $F = 1\varkappa-6$. Summing up we get $A+B+C+D+E+F+\varkappa = 7\varkappa+64$, which still equals $142+1\varkappa$. Therefore, $\varkappa = 13$. From this the remaining debts can be found.

The last problem in this section making use of several unknowns (fol. 254^v) asks for two numbers (say, P and Q) fulfilling the condition

$$P^2+Q^2-(P+Q) = 78 \ , \ \ PQ+(P+Q) = 39$$

(a reducible quartic). Stifel posits the first number to be 𝔷
and the second to be *A;* for convenience he represents
their sum by *B*. He proceeds in a way that has more to
do with *Elements* II or with square-grid geometry than
with algebra, using the diagram of Figure 10.

	39–1B	1z
1A	78+1B–1z	39–1B

Figure 10

The second condition gives him 𝔷A = 39–1B.
Thereby he can complete the square, etc.

So far, among the instances we have looked at, only
Antonio (and Pacioli's borrowed problem, above, p. 80)
used two unknowns in non-linear problems.[149] Two
things are to be observed in this connection. Firstly, that Stifel avoids using his new
formalism in non-linear *algebra*. Secondly (of great importance later on in our discussion),
that the geometric interpretation allows Stifel to take over from geometry the habit of
naming more than a
minimal set of unknowns
by letters. In a lettered
geometric diagram, *all*
occurring entities may
indeed be treated on an
equal footing.

	1A	1B
1r	240	660
1A	1Az	1380

Figure 11

The section on fols
292^r–301^r takes up the use
of several unknowns again.
The problems dealt with are
mostly linear – only two
are not. The first of these
(fol. 292^r) asks for three line segments, given the areas of the three rectangles they contain
pairwise. They are posited to be 𝔷, *A* and *B*. The solution follows from another square-grid
diagram (Figure 11). No algebra is used, and no equation formulated. As we see, the second
power of *A* is written as *A*ℨ, meaning "the *A*-kind of ℨ".

The last problem (fol. 300^v) – more difficult than the preceding ones (thus Stifel) and
serving to exhibit the potential of the technique – is inherently geometric but solved by
means of algebra. It deals with a rectangle with sides 12 and 14, subdivided into two
rectangles, the sum of whose diagonals is 28: By the Pythagorean theorem ab^2 is found
to be $12^2+(1\mathfrak{z})^2 = 1\mathfrak{z}$"+144, identified with 1*A*ℨ (i.e., (1*A*)^2). ad^2 is found in a similar
way to be 340–28𝔷+24ℨ, but also to be $(28–1A)^2 = 784–56A+1A$ℨ. This leads to 1*A*ℨ =

---

[149] From p. 33 we remember one seeming (but only seeming) exception from the Ottoboniano
Praticha: a linear problem solved by a procedure that leads to a reducible second-degree equation.

$56A+1\zeta-28\nu-444$. Since $1A\zeta = 1\zeta+144$, A can now be eliminated. The problem is thus of the second degree, but only in ν.

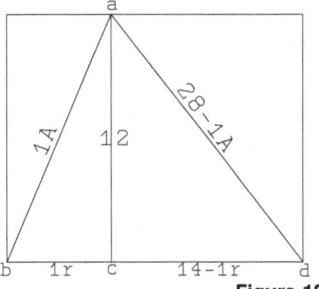

Figure 12

In [1553], Stifel prepared an "improved and much augmented" edition of Rudolff's *Coss*. He replaces Rudolff's notation for the second unknown by his own, and uses it in a number of problems that Rudolff had solved without using a second quantity. That teaches us nothing new.

An *Anhang* containing new problems does. 12 of 24 new problems (all of higher degree) operate with several unknowns. One of them may have been borrowed from Pacioli [1494: 148ᵛ], the question for two numbers fulfilling the conditions $mn = 96$, $m^2+n^2 = 292$, where Stifel makes the positions $m := \nu+A$, $n := \nu-A$.[150] Pacioli's parameters are different but his positions the same. The others I do not remember to have seen elsewhere. Two are solved by means of geometric diagrams similar to those we have seen, the others by means of algebra. All are quite sophisticated.

The potentially ambiguous notation for higher powers is now left behind: instead of $A\zeta$" Stifel here writes AA (etc.); in one problem (fol. 469ʳ), powers until k, ζA, νAA and AAA appear.

Immediate impact in Germany

Two works written in German land show (limited) influence from Stifel's new technique.[151] One is Mattheus Nefe's *Zwey newe Rechenbuecher* from [1565]. Nefe does not treat algebra in general (that falls outside the scope of standard *Rechenbücher*); on (fol. P iiiʳ), however, he gives a single example of *Regula quantitatis*. The words might be inspired by Rudolff – "without understanding this rule one can do little that is useful in the *coß*, and in others much less"; but he must also have known Stifel (directly or indirectly). His example is of type "all except each" and has four participants. In so far it is thus not innovative; but Nefe has seen that there is no reason to regard one unknown as primary; his names are therefore A, B, C and D.

Caspar Peucer's Latin *Logistice Regulae Arithmeticae, quam Cossam & Algebram quadratam vocant*, the second part of [Peucer 1556], was written and printed in Wittenberg. It was apparently meant for the higher Lutheran educational system, but does not seem to have been a conspicuous commercial success (neither official nor pirated reprints are known). Towards the end (fol. T viʳ) it contains a section *De radicibus secundis*, "on second

---

[150] The problem, as we have seen on p. 103, goes back to Antonio. Stifel, however, is not likely to have had access to Italian manuscripts.

[151] There may be more, of course. I have inspected some hundred 16th-century *Rechenbücher*, but there are many more; nor does "inspected" mean that have I read everything in detail.

roots", which initially refers to Rudolff, Cardano and Stifel. The notation is that of the *Arithmetica integra* ("1 *A*, id est 1 A$\mathfrak{z}$", etc.). In agreement with the Humanist program, some examples are drawn from Greek arithmetical epigrams (mathematical riddles originally meant to be solved without algebra, and thus by necessity elementary). One is of type "all except each". None are spectacular.

Mennher

Of much greater interest, and perhaps of decisive importance, is Mennher's *Arithmétique seconde* (above, p. 67). First in the second part – the algebra – comes a very orderly exposition of traditional *coß* with a single unknown. Then, toward the end (fol. O i^v), Mennher introduces the *règle de la quantité ou seconde radix*. The exposition is inspired by Stifel, but Mennher does not copy. His notation is the one used by Stifel in [1553]. Since not many books were around that presented the technique and apparently none treating it in depth we may safely assume that Mennher was working on his own. As we see, he is also not shy of speaking about his sources.

First (fol. O i^v) comes an "all less each" problem with four participants. It is similar to one in [Stifel 1553: 312^r]. The parameters are different, however; more important, so is the choice of the unknowns, for which reason Mennher cannot just proceed like Stifel. The third problem (fol. O ii^r) is similar to Stifel's two-number problem from [1544: 254^v]; the parameters are different, but the argument similar (including the use of a convenient though somewhat superfluous third unknown). Other two-number problems are similar to what we can find in [Stifel 1553] but solved differently – either by a better choice of the algebraic unknowns or by using a diagram instead of algebra; still others are without counterpart in Stifel.

All in all: Stifel took up the use of two unknowns in non-linear problems tentatively in [1544], and less tentatively in [1553]. In [1556] Mennher, though producing no striking further developments, took care that Stifel's technique could also reach French readers. In [1565: Ffi^r–Ggii^v] he was to present the *règle de la quantité* once again, by then expanding the treatment, though not much.

Antwerpen was an internationally well connected city, and Mennher was not forgotten. Descartes' friend and mentor Isaac Beeckman (on whom more below) possessed one of Mennher's arithmetics at his death – see [Beeckmann 1637x: B iv^v]; in a letter from around 1666, moreover, John Collins [ed. Beeley & Scriba 2005] mentions Mennher together with Viète and Viète's translator Jean-Louis Vaulezard (etc.) as good introductions to algebra.

Viète, as mentioned above (p. 78), only identifies predecessors and colleagues when he can censure them. Given his interest in spherical geometry, however, it is unlikely that he did not know at least what Mennher had written about that topic in [1564].

French writers after de la Roche

De la Roche's use of the *règle de la quantité* had no substantial impact on later French algebraic writers. The next French writer to deal with several unknowns is Peletier in [1554] (above, p. 42), who (as already mentioned) has not even heard about this predecessor.[152] Instead, his main inspiration even on this topic is Stifel. A section *Des racines secondes* (pp. 95–117) presents Stifel's system from the *Arithmetica integra* together with a number of examples, all except one of the first degree. The one which is not coincides (apart from the numerical parameters) with Stifel's first higher-degree problem. Peletier solves it by means of a geometric diagram, as already Stifel; in contrast to Stifel, however, Peletier refers explicitly to *Elements* II (namely, *Elements* II.4).

Jean Borrel (above, p. 43) borrows de la Roche's phrase *regula quantitatis* [Buteo 1559: 189]. He refers to Pacioli and de la Roche, leaving no doubt that he knew them, but to neither Stifel nor Peletier; even Borrel's habit, however, was to cite only those predecessors whom he could castigate. In any case, Borrel's are 1*A*, 1*B*, etc., apparently the same reduction of Stifel's (or Peletier's) system as effectuated a few years later by Nefe (in his notation for a single unknown, as we remember from p. 62, Borrel also changes what he had found in his model).

Since Borrel presents only 5 problems using the technique, all of the first degree (pp. 189–196, 357) he has no occasion to introduce higher power or products of the unknowns. He uses schemes for the operations on multiple equations, not too different from what had been done by Benedetto (whose work Borrel certainly did not know). There cannot have been much to learn from him for those who had studied Mennher's *Arithmetique seconde* (or even Peletier's *L'algèbre*).

Nor would there have been anything to learn *on this account* from Guillaume Gosselin's *De arte magna* from [1577] (above, p. 43). Toward the end of that book (fols 80^r–84^r) Gosselin deals with the topic of several unknowns under the heading *De quantitate absoluta*.[153] His names are the same as those of Borrel (that is, the same as those of Stifel apart from elimination of the *thing* as a primary unknown), and all his problems are of the first degree.

---

[152] De la Roche is not among those writers on algebra whom Peletier had either read or knew about without having seen their works. In the Latin version of the work from [1560] de la Roche does turn up in the unpaginated preface, but only as somebody who has written on algebra.

[153] This name may have been taken over from Pedro Nuñez [1567: 224^v], a book referred to by Gosselin. Nuñez operates with *cosa* and *quantidad*, apparently borrowing from Pacioli, to whom Nuñez refers (fol. 225^v).

Chapter VI. The transition to incipient modern algebra

In the preceding chapters we have traced the development of a number of characteristics of the algebra that emerged in 17th-century France. A development with stops and goes, with false starts and long stagnations – rarely involving deliberate explorations, and when, then often not emulated by the following generations.

More false trails – trails leading nowhere *for the moment* – could have been included. Fibonacci did extensive work on the Euclidean theory of irrationals applied to numbers and their roots, and so did Benedetto, Pacioli, Stifel and Stevin. This theory played no major role in the emergence new algebra, however, but had to be taken up again in other contexts and in later times. It is not quite as much a dead end as the geometric proofs for the basic algebraic cases, which were not predestined to be ever integrated in new developments, being after all mathematically trivial in the new context; but almost.

Evidently, even the characteristics that we have followed in Chapters II–V did not *by necessity* end up integrated in a new whole, although our starting point was that they did end up being integrated. We have argued so far that the early 17th had flour, eggs, butter and sugar at disposition. That did not automatically produce cake. Was the cake the outcome of a deliberate decision to make it? And whether this be the case or not, which was the oven where it happened?

It happened in France, as we know. But from the story as told here, nothing in the internal development of mathematics pointed to France. In order to understand the role of France, we have to step outside the closed space of mathematical techniques.

Mathematical cultures

In recent centuries algebra, even higher-degree algebra (not to speak of offsprings like the infinitesimal calculus) are essential to many applications of mathematics. That was not the situation during the centuries where abbacus algebra and *coß* flourished, nor in the 17th century. So, why were brilliant mathematical minds engaged in developing algebra, and why did less brilliant minds try to show that they understood it? The answers for the late Middle Ages and the 17th century are deceptively similar.

The *raison-d'être* and format of abbacus algebra

Abbacus masters were members of a liberal profession in a market economy, as pointed out on p. 11. They competed with each other, either for students or for positions at municipally financed abbacus schools. There, ability to solve problems proposed by the

J. Høyrup, *Explorations and False Trails*, SpringerBriefs in History of Science and Technology, https://doi.org/10.1007/978-3-031-48158-1_6

adversary at a competition or the aptitude to propose problems the adversary could not solve were useful. One effect of that was the spread of false rules claiming to provide solutions to irreducible cubics and quartics [see, e.g,. [Høyrup 2009: 50–52]; another, of greater general importance, was its synergy with the aim of the teaching in the abbacus school. Even school students had to be trained in solving problems – simple problems in their case. Competition as well as abbacus-school teaching pushed toward seeing *problem solution*, as the general format in which mathematical knowledge should be formulated.

Fibonacci, Benedetto, Pacioli (and a few others) took up the study of the Euclidean irrationals, and there the format was evidently different. It was thus *possible* for abbacus writers to formulate knowledge in a format that approached the exposition of theory (though with scant proofs). But that style did not spread; when it was realized and formulated by Pacioli [1494: 150^f] and others from his times onward that equations involving three powers can only be solved by general rules if the three powers are equidistant in the sequence of powers, this is taken note of as an empirical fact; in the *coß*, this insight is *used* when the many rules are reduced to eight.

Rechenmeister culture was not intellectually competitive as abbacus culture had been, or at least not to the same extent (some wry or ironical remarks indicate public intellectual scorn was not totally absent); those *Rechenmeister* who wrote books competed *on the book market.* But they had no incentive to change the inherited preferred format for mathematical knowledge – their audience still wanted to learn how to solve practical and practically looking problems; in consequence the *Rechenmeister* did not change the preferred format.

Scholastic mathematics

The situation of mathematics in the Latin tradition had been totally different. Its main constituents were these:
– since the later 10th and 11th centuries, Boethian arithmetic and music theory, taught in the cathedral schools and later in the earlier stage of the arts faculty curriculum;
– from around the same time, the use of the "Gerbert abacus", taught in the cathedral schools – but then replaced by
– algorism, teaching the use of the Hindu-Arabic numerals, mainly aiming at astronomical calculation;
– lectures on the *Elements*, on Witelo's *Perspectiva communis*, and other works.
The difficult categorization and naming of the Boethian ratios were trained by means of a board game *rithmomachia*, and probably also by repetition; the training of the Gerbert abacus and algorithm certainly made use of more than the few illustrative examples contained in the treatises, but still examples fitting within the framework of these.

Lectures would be followed up by disputations, and disputations about mathematical lectures might easily be concerned with metamathematical discussions about the status of objects of the lectures. That can be seen in the written emulations of disputations, the collections of *quaestiones*. The only way to dispute about a correctly performed Euclidean proof is indeed to challenge its foundations.

The disputations were agonistic by definition, but like the *quaestiones* they aimed at finding a decision, not at formulating new mathematical knowledge. The format for expressing mathematical knowledge remained that of the mathematical textbooks – the *Elements*, Jordanus's *Liber philotegni*, and more books joining them as time went on.[154] Characteristically, even Jordanus's *De numeris datis*, a domestication of Arabic algebra emulating the Euclidean *Data*, reformulates the Arabic problem solutions as theorems. The format of scholastic mathematics was *theory* expressed in *theorems*.

That format still governs Stifel's *Arithmetica integra*. This is the reason he has as his main aim to develop *a technique* for the use of several unknowns – his problem solutions are *illustrations*, not a primary aim. The format of the re-edition of Rudolff's *Coss*, in contrast, is the problem, and here, the purpose of the *Anhang* is to show that Stifel can do even better than Rudolff. This agonistic context invites the expansion of the system.

Humanist mathematics

"Humanist mathematics" – is that not a contradiction in terms (not in our irrelevant perspective but in that of the Renaissance, where "Humanist" points to the Humanist movement as embodied by Francesco Petrarca, Giovanni Boccaccio, Marsilio Ficino and their kind)?

Until around 1450 it certainly was. Then some interaction between engineers and architects like Brunelleschi with the Humanist current took off. These "higher artisans" (not really including the abbacus masters, though the Florentine encyclopedias exhibit some affinity, cf. note 84) tended to see themselves, and to be seen as "the Archimedes of our time", reflecting the Humanist understanding of Archimedes as an engineer serving his city and king, in the image of the court mathematicians of the epoch. The decisive jump, however, took place after 1500.

Humanism had always been concentrated on the "civically useful", originally literary and rhetorical service to the city elite or the city tyrant. By 1500 it had become clear that Latin letters did not suffice to protect the city walls against the French artillery, nor to plan the great drainage projects making land fit for agriculture (Bombelli was engaged in one). Nor would they help the Spanish and Portuguese Crowns to make themselves masters of as much of the world as they could grasp. Fortification, engineering and transoceanic navigation were all in need of mathematics, and for the Humanists mathematics had to be Greek mathematics to the extent this was possible (sometimes beyond the possible, as when mathematicians from Regiomontanus to Clavius ascribed algebra to Diophantos).

So, around 1500, a few Humanists started collecting and publishing or translating Greek

---

[154] This is not to say that there were regular lectures on all of these. The *Liber de triangulis Jordani* seems to be a student *reportatio* of lectures [Høyrup 1988:347–350]. This shows that lectures were held on the *Liber philotegni;* but it may well have happened only once, Jordanus himself being the likely lecturer. Similarly in other cases.

mathematical texts. This, we may observe in passing, is the time when "Northern Humanism" took off, the Humanism of Erasmus and Thomas More and soon of Melanchthon – the Humanism based exclusively on texts, while that of Italy had had Antiquity before its eyes though in ruin. 1500 is also close to the end of what historians of art consider the Renaissance – after 1520 they mostly prefer to speak about Mannerism, and soon Baroque.

Giorgio Valla's posthumous *De expetendis et fugiendis rebus* from [1501] contained Euclidean excerpts translated directly from the Greek, and also some Archimedes through Eutocios's commentary. Bartolomeo Zambertis' translation of the *Elements* and minor Euclidean works from an inferior Greek manuscript appeared in 1505. These volumes provided versions with "Humanist credentials" but not much that went beyond what was known in the Latin tradition.

Around 1500 (the beginning of French Humanism), Lefèvre d'Étaples started mathematical publishing. His perspective was Neoplatonic, but his mathematics was that of the scholastic tradition (Jordanus, epitomes of Boethius, *rithmomachia*, the *Elements*).

New old mathematical knowledge appeared from 1533 onward. That year saw Grynaeus's edition of the Greek Euclid with Proclos's commentary. The *editio princeps* of Pappos's *Collection* appeared in Basel in 1538, that of Archimedes in Basel in 1543; Memmo's Latin edition of books I–IV of Apollonios's *Conics* appeared in 1537 (Commandino's in 1566); Xylander's Latin translation of Diophantos, as mentioned, was published in 1575. "Humanist mathematics" thus took its beginning around 1530–1540, two centuries after Humanism proper.

Agonistic mathematics, once again

As we have noticed, many French algebraic writers after de la Roche and until Viète did not cite predecessors except in order to show themselves superior (Peletier and Gosselin being exceptions). This is quite different from the general absence of citations in the *Liber abbaci* or Pacioli's *Summa*,[155] and is evidence of agonistic behaviours (it would be misleading to speak about a *return* of these, since there is no inherent connection to the competitive confrontations of the abbacus masters).

Competition based on mathematical competence was no mere internal characteristic of the mathematical ambience. This is revealed by a familiar story [Busard 1976: 22]:

Viète's mathematical reputation was already considerable when the ambassador from the Netherlands remarked to Henry IV that France did not possess any geometricians capable of solving a problem propounded in 1593 by Adrian Romanus [van Roomen] to all

---

[155] Pacioli. it is true, states in the beginning of his unpaginated *Summario* that most of the *Summa* have been taken from Euclid, Boethius, Fibonacci, Jordanus, Blasius of Parma, Sacrobosco and Prosdocimo de' Beldomandi. Apart from Euclid and Fibonacci, however, this list of Latin writers has nothing to do with his real sources. It constitutes an oath of fealty, exactly the opposite of an attempt of the author to claim he had done better than these luminaries.

mathematicians and that required the solution of a forty-fifth-degree equation. The king thereupon summoned Viète and informed him of the challenge. Viète saw that the equation was satisfied by the chord of a circle (of unit radius) that subtends an angle $2\pi/45$ at the center. In a few minutes he gave the king one solution of the problem written in pencil and, the next day, twenty-two more.

In the Italian Hight Renaissance, good Latin style had been a diplomatic weapon for Florence and other Italian Renaissance city states; that is the reason they appreciated Humanists.[156] Now, two centuries later, it was the turn of mathematical prowess to become an *affaire d'état*.[157] As further illustrated by our story, this prowess was put into play by the ability to *solve problems*. No longer sophisticated versions of recreational classics – "purchase of a horse", "men finding a purse", etc. In the van-Roomen–Viète case, as we see, it had to do with new developments of trigonometry, but more broadly it was defined by the new Humanist mathematics.

The kind of problems by which late-16th-century mathematicians challenged each other is reflected in Viète's *Variorum de rebus mathematicis responsorum liber VIII*, "Book 8 of various responses about mathematical matters" [1593a]:
– two intermediate proportionals;
– squaring and rectification of the circle and of circular segments, using Archimedean spirals and the quadratrix;
– construction of a regular heptagon;
– lunules; etc.
In the end Viète deals with spherical trigonometry, a topic that had his special interest; this is the only topic that points to broader practices (astronomy and navigation). Soon, Pierre de Fermat, Gilles de Roberval and of course Descartes were to widen the horizon, taking up not only further areas of Greek mathematics but also, for instance, the geometry of kinematics. Geometry, however, was core as well as periphery.[158]

---

[156] The Duke of Milan "was often heard to say that he was not damaged as much by a thousand mounted Florentine warriors as by Coluccio Salutati's style" [Gragg 1927: x]. Salutati was the Humanist Chancellor of Florence from 1375 to his death in 1406.

[157] Not only mathematical prowess. The princely interest in mathematical splendour has a close parallel in the phenomenon of 16th-century "courtly science" as described by William Eamon [1991: 35], characterized (along with interest in the occult) by

fascination with and the display of meraviglia, which is best seen in the princely gardens and cabinets of curiosities [...] symbolically demonstrating the prince's dominion over the entire natural and artificial world. Carved gems, watches, antiques, mummies and mechanical contrivances were displayed side by side with fossils, shells, giant's teeth, unicorn's horns, and exotic specimens from the New World.

This was certainly different from the incipient new natural history; but it provided a substrate and foundation for patronage of that field.

[158] Thanks to Fermat, this sweeping generalization is not *quite* true. Chapter XCVIII of John Wallis's

An apparently minor leap with immense consequences

Neither the "return" to problem-solving as the central manifestation of mathematical proficiency nor the restoration of Greek geometry produced the new algebra directly, neither singly or in combination. Neither, indeed, had to do with algebra. Their importance derives from the decision of Viète and Descartes to *apply* algebra in geometric problem-solving.[159] Algebra, indeed, had never been a theory, but always existed as a tool for solving problems, albeit mostly non-geometric problems.

Both speak in derogatory terms about the algebra they had inherited. Viète speaks of algebra as "a new art, or rather so old and so defiled and polluted by barbarians that I have found it necessary to bring it into, and invent, a completely new form" [1591a: 2^v]. In the *Discours de la méthode*, Descartes [1637: 19] speaks about existing algebra as "a confused and obscure art that puts the mind in difficulty instead of a science that cultivates it".

Both, on the other hand, were chasing a tool that would allow them to solve all problems. Viète famously closes his *Isagoge* [1591a: 9^r], the introductory part of his reconstruction of algebra, with the motto *nullum non problema solvere*, "to leave no problem unsolved" (revealing that he cannot imagine other kinds of problems than those pertaining to geometry). In a letter to Isaac Beeckman from 1619, Descartes expresses the somewhat better defined ambition to solve all problems "dealing with any kind of quantities, discrete as well as continuous", by means of curves corresponding to higher-degree equations [ed. Adam & Tannery 1897: 157].

But which was the algebra they were speaking about?

Algebra as Descartes knew it

Descartes frequented the Jesuit *collège* La Flèche from he was 10 until he was 18 (1606–1614). This is where, as a youngster, as we read in *Discours de la méthode* [Descartes 1637: 18], he had read some

> logic and, among the mathematics, the analysis of the geometers and algebra, three arts or sciences which seemed to promise something for my purpose.

Treatise on Algebra [1685: 363–371] thus takes as its starting point a number-theoretical problem proposed by Fermat in 1657 "as a challenge to all the mathematicians of *Europe*". It is already presented, along with other number-theoretical problems linked to Fermat, in Wallis's *Commercium epistolicum* from [1658: 34]. Apart from Fermat, van Schooten and Wallis himself, the correspondents are *dilettanti*, whose mathematical ability equalled their (not supreme) nobility.

[159] This is quite different from what Nuñez did to algebra and geometry in [1567: 227^v–331^r]: Wishing to promote algebra, he illustrated the potential of the technique by showing how it could solve geometric problems chosen for the purpose (and mostly traditional). This did not ask for any change of algebra as he had received it, and produced none. Cf. [Høyrup 2002].

Algebra he had been taught on the basis of Clavius's textbook [1608], a *coß* in German style though in Latin, much in debt to Stifel. If the teacher had gone until p. 72, Descartes may also already have learned about the use of several unknowns, in Stifel's notation from 1544 (not 1553), but expanded also to negative powers.

Thinking back (when he had created his own notations in the *Géométrie*) Descartes thought (as just quoted) of Clavius algebra as "a confused and obscure art that puts the mind in difficulty instead of a science that cultivates it". Yet he still uses Clavius's cossic notation in the letter to Beeckman from 1619 in which he expresses his ambitions to solve all problems "dealing with any kind of quantities" – by the way without betraying the least trace of familiarity with several unknowns.

In 1628 Descartes met Beeckman again; in a note [ed. Adam & Tannery 1908: 334] Beeckman reports that Descartes has told him to have invented a general algebra where all the *notae cossicae*, the "cossic characters", are represented by lines (as they were to be in the *Géométrie*). But the first four powers are still written $\curlyvee$, ζ, $c\!\!/$ and $\zeta\zeta$; in a simple algebraic calculation given as example, binomials are still added in a scheme.

By then, if not before, Beeckman may have introduced Descartes to Mennher, or Descartes may have discovered Mennher in Beeckman's library. In any case there is still no trace of influences other than the *coß*, mediated either through Clavius or perhaps through Mennher. And there is certainly not the slightest suggestion that Descartes should have read Viète.[160]

In the *Géométrie* Descartes [1637: 398–400] also knows Cardano's *Ars magna*, but that seems to be a recent acquaintance; in any case it will not have helped him to bring about the new "general algebra".

Viète's "defiled and polluted" algebra

Descartes does not himself identify the algebra that "puts the mind in difficulty"; *we* can do so because we know his school book and because we can recognize its ways in his interactions with Beeckman. If we had possessed only the *Géométrie* we would have been at a loss.

So we seem to be with Viète. His terms for the powers are obviously borrowed from Xylander, perhaps influenced by Gosselin. He mentions Diophantos and other Greek authors (without taking over more than the names for powers and such terms as *analysis* and *zetetic*), but he is almost mute regarding writers on algebra – he mentions Cardano's *Practica arithmeticae* [1539] twice, but that is all. The contents of his writings confirm that he had "found it necessary to bring it into, and invent, a completely new form". We

---

[160] In a long patriotic rant, Nathaniel Hammond [1742: xvii] claims that Descartes had learned from [Harriot 1632] (and that the *Géométrie* was published anonymously in 1637!). The fancy can be rejected already for chronological reasons: as shown by Christian van Randenborgh [2012: 225], van Schooten was acquainted with the manuscript already in 1632.

may say that his solutions of the third- and fourth-degree equations show that he knew the *Ars magna* – cf. [Witmer 1983: 4 n.7]. Beyond that, many of the problems in the *Zetetica* [Viète 1591b] are reformulations of what is found in numerous abbacus algebras, but precisely because it is found in many books and the formulations agree with none of them, we cannot say *which* books he brought into a new form. His interest in spherical geometry suggest he *could* have read Mennher; for the same reason. he could have looked at Nuñez. From the former he might have learned about the use of several unknowns (but Peletier and Borrel are alternative possibilities, as even the *Arithmetica integra*); from the latter not.

Receiving abstract coefficients as a gift not asked for

Viète also keeps his mouth shut regarding the introduction of abstract coefficients. Descartes is less taciturn. Once we have understood why Descartes did as he did, we may see how Viète fits within the same picture.

In 1628, Descartes' ambition had been to solve all problems about continuous as well as discrete quantities. In the *Géométrie*, the discrete quantities have disappeared – the *Géométrie* is, precisely, about geometry and not about number theory, and about how to solve geometric problems.

The central subject is "Pappos's problem" – see, for instance, [Bos 2001: 272–274]. That, however, is immaterial to what we are discussing here. Decisive is that the problem is formulated around a lettered diagram. In a lettered geometric diagram, as observed on p. 108, *all* occurring entities may indeed be treated on an equal footing.

When geometers of the time (indeed geometers since Euclid) wished to solve a problem, they looked at or drew a lettered diagram – a paradigmatic example displaying its structure.[161] What to do if they wanted to solve it by means of algebra? Descartes [1637: 300] tells us:

> When wishing to solve some problem, one should first look at it as already solved, and give a name to all the lines that seem to be needed in order to construct it, those that are unknown as well as the others. Then, without making any difference between these known and unknown lines, one should run through the difficulty according to the order it shows,

---

[161] It is often overlooked that even the diagrams used in Euclid's proofs are paradigmatic examples. The propositions are abstract, but the proofs based on a specific configuration. If our misconception of the nature of Greek mathematics should prevent us from seeing that, we may look at *Elements* V.1 [trans. Heath 1926: 138]:

> If there be any number of magnitudes whatever which are, respectively, equimultiples of any magnitude equal in multitude, then, whatever multiple one of the magnitude is of one, that multiple also will all be of all.

In *the proof*, Euclid takes the "any number of magnitudes" to be 2, and similarly, the "equimultiples ... equal in multitude" to be 2. Even here the proof is performed on a paradigmatic example – neither more nor less general than the solutions offered by abbacus masters for their horse problems.

the most natural of all, in which way they depend mutually on each other, until the point where one has found a way to express one and the same quantity in two ways: which is called an equation.

As already Stifel, Descartes does not look for a minimal set of unknowns – everything that seems to play a role gets a name. Once the idea to apply algebra with several unknowns to geometric problems is there, quasi-abstract names (that is, not specified numbers but names linked to specific entities appearing in a diagram) would be used for everything pertinent with no distinction between what would turn up as coefficients in the equations and what would turn up as unknowns.

Slightly later, Descartes introduces (but does not explain) the principle to use letters from the end of the alphabet (first z, then if needed also y, and then x) for the unknown magnitudes and letters from its beginning for those that are known – the quasi-abstract coefficients.

We should take note of the words "all the lines that seem to be needed". A traditional lettered diagram would mostly give letter-names to *points*, not to the segments. But algebra applies to quantities, and these are the segments. Expressing the equations by means of two letters defining each of these would be cumbersome, and accordingly Descartes (as we have seen it in Stifel, and as we shall see it in Viète) tacitly changes the way lettering is made. The quasi-abstract coefficients presented themselves as a gift that could not be refused; but as such gifts usually do, this one came together with conditions.

Geometry and algebra with one unknown

Applying algebra with an unlimited number of unknowns to geometry led Descartes quite naturally to those quasi-abstract coefficients that were to become fully abstract once the new algebra left its native geometric soil behind. We may compare to what an equally talented geometer had to do when applying algebra with a single unknown to intricate geometry.

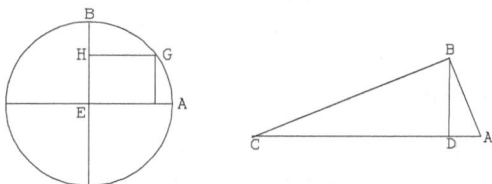

Figure 13

That geometer is 'Umar al-Khayyāmī, certainly not inferior to Viète and Descartes as a mathematician. A small treatise of his [ed. trans. Rashed & Djebbar 1981: I, 73–90] deals with a particular partition of a circular arc. The arc AB (left in Figure 13) is to be divided at G in such a way that $AE : GH = EH : HB$. A long analysis reduces this to the finding of a right-angled triangle ABC (right in the diagram), with height BD, in which $AB+BD = AC$. In order to apply *his* algebra with only one unknown, al-Khayyāmī needs to posit that $AD = 10$; that leads him to an equation whose coefficients are numerically fixed. Descartes (as well as Viète) would have posited AD to be, for instance, b, which would automatically (though obviously after as much calculation as made by al-Khayyāmī) have produced an equation with quasi-abstract coefficients.

Viète's abstract coefficients

Viète tells nothing directly about how he gets to the abstract coefficients – which, since he sets out in [1591a] and [1591b] with his new algebra and not its applications to geometry, are really abstract. He even conceptualizes the innovation [1591a: 5^r], distinguishing between *logistica numerosa*, where the coefficients are numbers, and *logistica speciosa*, where the *species* or *forms* of things are shown.[162]

In any case, there is little doubt (if we combine his motto *nulla non problema solvere* with his actual interests outside algebra) that his motive for creating a new algebra was its application to geometric problem-solving. If we look at what he does when he applies his new technique to geometric problems – for instance, in the *Zetetica* [Viète 1591b: 14^v–15^r] we also observe that letter names are given to segments, not to points. In contrast, in the *Supplementum geometriae* [Viète 1593b], where all proofs are based on proportion theory and so-called "geometric algebra", points and not segments carry letter names.

In chapter V of the *Isagoge* (fol. 7^r), "On the laws of zetetics", Viète describes the procedure to be used when an algebraic solution is aimed at:

> Magnitudes, those which are known as well as those which are asked for, should be combined and compared, adding, subtracting, multiplying and dividing, always observing the law of homogeneity.

All in all, Viète's road toward the abstract coefficients (in the latter description, with its implicit basis in a diagram, quasi-abstract) appears to have been very similar to that of Descartes. Even Viète received them as a gift not asked for but coming by necessity out of the application of algebra with an unlimited number of unknowns to complicated geometric problems – he did not "invent" them out of the blue.

Descartes' other innovation

The introduction of quasi-abstract coefficients (considered fully abstract by almost everybody) is usually seen as the watershed between pre-modern and modern algebra. Also fundamental, as claimed on p. 4, is the establishment of the general parenthesis function. This other pre-requisite for the emergence of modern algebra was absent from Viète's work but so essential that his interpreters had to read it into them – see p. 72. What they read into them, however, would not have sufficed, for instance, for Euler's purposes.

---

[162] This reference to "species or forms of things" seems to be borrowed from scholastic Aristotelian philosophy.

What about Descartes?

Descartes makes use of three different parentheses, all of them with us today, or almost. The first is the old fraction line, the second the extended root sign $\sqrt{}$, which allows nesting

$$\sqrt{-\tfrac{1}{2}+\sqrt{\tfrac{1}{4}aa+bb}}$$

[Descartes 1637: 303]. It is not different in idea from what Bombelli had used, but it happens to have been conserved. The third, though looking different, might be taken for an immature fore-runner of our present-day all-purpose parenthesis (but see imminently).

The first occurrence is this (left how it looks in the collected works [ed. Adam & Tannery 1902: 398]; right, the original print [Descartes 1637: 325]:

$$yy \infty \dfrac{\left.\begin{array}{l}-dekzz\\+efglz\end{array}\right\}\, y \left.\begin{array}{l}-dezzx\\-efgzx\\+bcgzx\end{array}\right\}\, y \left.\begin{array}{l}+bcfglx\\-bcfgxx\end{array}\right\}}{ezzz-cgzz},$$

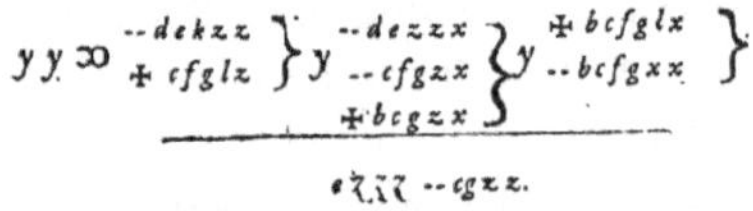

As we see, the parentheses are not enclosed in pairs of brackets; they are written vertically and kept together by a brace to the right. That should be immaterial, however, as long as they are unambiguous. It is also irrelevant that the brace had been used by Viète with a different function; not quite as irrelevant, perhaps, is Vaulezard's parenthesis-like reinterpretation of Viète's notation (above, p. 71). In principle it should allow nesting, but Descartes never tries – his applies this third kind of parenthesis rarely, and only to express a composite coefficient. On a few occasions he omits the brace but indicates the structure by using a reduced type size for the elements of the parenthesis (unfortunately rendered badly in the collected works – see Figure 14). It would hardly be possible

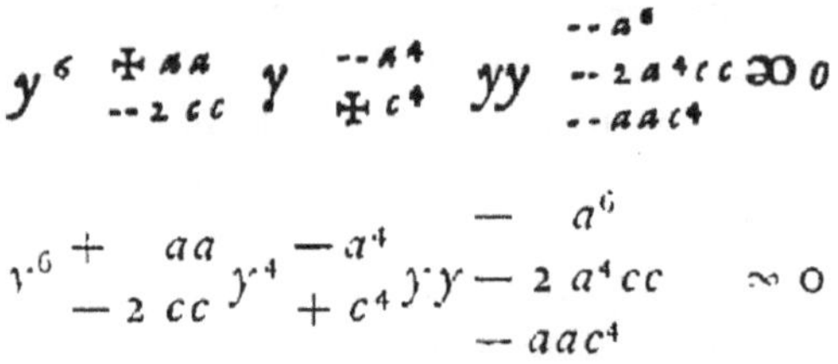

Figure 14. Descartes' brace-less parenthesis [top Descartes 1637: 384; bottom Adam & Tannery 1902: 458]

to use the notation to express a power of a polynomial, or the product of two polynomials. All in all, what Descartes has invented here is a third kind of special-purpose parenthesis and not the all-purpose parenthesis. This time he has not received a gift not asked for, the task has forced him to invent, not too different from what had happened to Antonio, Benedetto and Cardano. As often with a first invention under such circumstances, the result is somewhat clumsy.

That his invention had greater consequences than Antonio's nested roots or Benedetto's linear algebra with multiple unknowns depended on Descartes's readers. Even they had sparse use for a general parenthesis, but occasionally they encountered the need to widen the use of what they had learned from Descartes.

Newton and Wallis

Two readers who widened the parenthesis idea are Isaac Newton and Wallis. There will certainly be others, but these are the two I looked at for the purpose.

Newton's *Arithmetica universalis* was published in [1707], but it is based on algebra lectures he held during the 1670s.[163] Like Descartes' *Géometrie* it gets most of what it needs from the classical special-purpose parentheses, the fraction line and the extended root sign. Beyond that, it generally uses the vinculum, a stroke above, to indicate a parenthesis. It looks like the extension of the root sign without the root sign itself, and that is certainly how the printer made it (his extended root sign is indeed made in two unconnected parts). Whether Newton got his idea from there can probably not be known with certainty (nor is it of major importance except if we take up the psychology of creativity).[164] Among other things he uses it for the products of two polynomials (see Figure 15, top). It can be justly characterized as defining a general parenthesis.

Newton also knows Descartes brace-delimited composite coefficient, and uses it when needed; an example is shown in Figure 15, middle. As we see, Newton uses the variant where the brace is omitted. That may be because he uses the brace for a different purpose (see Figure 15, bottom), namely exactly as Viète had done. In principle he might thus have read and understood Viète – but since the "Viète" that was widely known was the one we find in [van Schooten 1646], this is unlikely. Others at the time use the brace in a similar way, and Newton probably followed their customs.

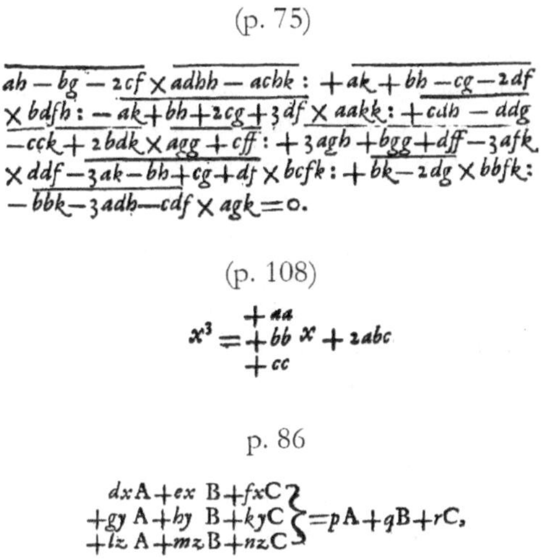

(p. 75)

$$\overline{ab - \overline{bg} - 2cf} \times \overline{adbh - achk} : +\overline{ak} + \overline{bh} - \overline{cg} - 2df$$
$$\times \overline{bdfh} : -\overline{ak} + \overline{bh} + 2cg + 3df \times \overline{aakk} : +\overline{cdh} - ddg$$
$$-\overline{cck} + 2bdk \times \overline{agg} + cff : +3agh + \overline{bgg} + dff - 3afk$$
$$\times \overline{ddf} - 3ak - bh + cg + df \times bcfk : +\overline{bk} - 2dg \times bbfk :$$
$$-\overline{bbk} - 3adh - cdf \times agk = 0.$$

(p. 108)

$$x^3 = {+aa \atop +bb \atop +cc} \, x + 2abc$$

p. 86

$$\begin{matrix} dx\,A + ex\ B + fx\,C \\ +gy\,A + hy\ B + ky\,C \\ +lz\,A + mz\,B + nz\,C \end{matrix} \Big\} = p\,A + q\,B + r\,C,$$

Figure 15

---

[163] More could be learned about *Newton's thought* from his manuscripts, but not about his contributions to historical development. According to what I have seen in them they tell nothing important on the present account beyond what already follows from scrutiny of the *Arithmetica universalis*. In the published work, just as in the manuscripts, Newton appears to have understood that Descartes' brace is just another specific-purpose parenthesis, and he only uses it as such.

[164] There is some evidence that Newton saw the connection. In *Analysis per Quantitatum Series, Fluxiones ac Differentias* [1711: 53 and *passim*] he speaks of composite radicands as *vinculum radicis*. The manuscript seems to be from 1671 [Westfall 1671: 135] and thus to predate the algebra lectures.

Wallis went further in his extensive *Treatise on Algebra* from [1685].[165] Even he uses the classical special-purpose parentheses profusely, and rarely needs more. On one point, however, he not only goes further but much further, namely on pp. 332–333. The context is quadratures and development in infinite series. Still geometry of conic sections, for sure, yet not the kind Viète, Fermat and Descartes had dealt with but an early step in infinitesimal calculus. Here, parentheses marked by vincula are raised to integer or broken powers (the latter another radical innovation), such as $\overline{d+e}|^{\frac{3}{4}}$.[166]

A look at writings treating of infinitesimal analysis from the next two decades, for example Guillaume François Antoine de l'Hospital's *Analyse des infiniment petits* from [1696] and the Gottfried Wilhelm Leibniz-Johann Bernoulli correspondence [ed. Bousquet 1745] shows that vincula as well as parentheses contained by round brackets abound – together of course with the traditional special-purpose parentheses, that is, the fraction line and the extended root sign.

The truly general parenthesis

Infinitesimal calculus could not do without the all-purpose parenthesis. After the isolated experiments of the late 17th century, 18th century higher mathematics adopted it to the full, as a tool so obvious that there was no need to notice it (the quip that fish do not know about water applies here). That opened the road to such fabulous calculations (to mention but this example) as Euler's development of the infinite fractional product [1748: I, 257]

$$\frac{1}{(1-xz)(1-x^2z)(1-x^3z)(1-x^4z)(1-x^5z)\ \&c.,}$$

as the sum

---

[165] Other works of his would illustrate his very informal ways, for instance, his insertion of rhetorical explanatory parentheses inside mathematical expressions – like Cardano, Wallis seems to have been a great mathematician not because of but in spite of his notations. In the present perspective, what I have observed in these other works was unimportant.

[166] The fractional exponents are introduced on p. 288:

> Understanding, by the Exponent of the Side, Square, Cube, &c the whole Numbers 1,2, 3, &c, and of the Roots Quadratick, Cubick, &c, the Fracted Numbers $\frac{1}{2}$, $\frac{1}{3}$, &c; and of their Compounds, the Square Root of Cubes, the Cubick Root of Biquadrates,&c. $\frac{3}{2}$, $\frac{4}{3}$ &c; and of their Reciprocals; –1, –2, –3, $-\frac{1}{2}$, $-\frac{1}{3}$, $-\frac{3}{2}$, $-\frac{4}{3}$, &c.

Elsewhere, Wallis uses "exponent" in the sense of "identifier" (as Peletier): as a designation for the fraction $^p/_q$ corresponding to a ratio $p : q$ (in the preceding tradition known as its "denomination"); and for the factor of increase in a geometric progression. See [Wallis 1585: 78, 89].

$$
\begin{aligned}
&1{+}z\ (x\ {+}x^2{+}\ x^3\ {+}\ x^4\ {+}\ x^5\ {+}\ x^6\ {+}\ x^7\ {+}\ x^8\ {+}\ \ x^9\ {+}\&\text{c.})\\
&{+}z^2(x^2{+}x^3{+}2x^4\ {+}2x^5\ {+}3x^6\ {+}3x^7\ {+}4x^8\ {+}\ 4x^9\ {+}\ 5x^{10}{+}\&\text{c})\\
&{+}z^3(x^3{+}x^4{+}2x^5\ {+}3x^6\ {+}4x^7\ {+}5x^8\ {+}\ 7x^9\ {+}\ 8x^{10}{+}10x^{11}{+}\&\text{c.})\\
&{+}z^4(x^4{+}x^5{+}\ 2x^6\ {+}3x^7\ {+}5x^8\ {+}6x^9\ {+}9x^{10}{+}11x^{11}{+}15x^{12}{+}\&\text{c.})\\
&{+}z^5\ (x^5{+}x^6{+}\ 2x^7\ {+}3x^8\ {+}5x^9\ {+}7x^{10}{+}10x^{11}{+}13x^{12}{+}18x^{13}{+}\&\text{c.})\\
&{+}z^6(x^6{+}x^7{+}\ 2x^8\ {+}3x^9\ {+}5x^{10}{+}7x^{11}{+}11x^{12}{+}14x^{13}{+}20x^{14}{+}\&\text{c.})\\
&{+}z^7(x^7{+}x^8{+}2x^9\ {+}3x^{10}{+}5x^{11}{+}7x^{12}{+}11x^{13}{+}15x^{14}{+}21x^{15}{+}\&\text{c})\\
&{+}z^8(x^8{+}x^9{+}2x^{10}{+}3x^{11}{+}5x^{12}{+}7x^{13}{+}11x^{14}{+}15x^{15}{+}22x^{16}{+}\&\text{c.})\\
&\qquad\qquad\qquad\qquad\&\text{c.,}
\end{aligned}
$$

After that, though hesitantly, even ordinary and elementary algebra received the general parenthesis as a gift – but only gradually, and from its own offspring.[167]

Once the quasi-abstract coefficients had been introduced, they had been adopted systematically by every participant in the development of the new mathematics of the 17th century – Harriot, Fermat, etc. They were the *sine qua non* of participation in that endeavour.

The general parenthesis, instead, was the *sine qua non* of 18th-century mathematics. The story promised in the Introduction is thereby finished.

---

[167] In [1726: 32*f*], Jean-Pierre de Crouzas uses the vinculum as Newton had done in the *Arithmetica universalis* for the multiplication of two polynomials, and on p. 283 in order to explain the universal root (after that the extended root sign with nesting is used). On a few occasions the vinculum is used where it is algebraically superfluous, to delimit a complete expression (p. 412 and *passim*), and on p. 469 when the square on a line segment named with two letters is expressed – none of these are algebraic uses. On p. 481, a brace is used in Viète's manner. In a book of almost 500 pages, this must be said to be a modest use of the general parenthesis.

In [1742], Hammond published *The Elements of Algebra* intended "for beginners" and to be "as useful as possible to the Publick schools" (thus the preface, p. v) and mainly based on [Harriot 1631]. It is thorough, everything meticulously explained with many examples. Mostly these are constructed in such a way that only the classical special-purpose parentheses are needed (for the composite radicand it uses the extended root sign instead of Harriot's ambiguous notation). On p. 261, however, and again on pp. 263 and 268, Hammond cannot avoid the vinculum (explaining on p. 261 as well as 278 what it effectuates).

At the exams of St. John's College in 1797 and 1800 (that is, also in Eton-Oxbridge context), the students were expected to understand and solve this equation [Rotherham 852: 4]:

$$
\frac{123+41\sqrt{x}}{5\sqrt{x}-x} = \frac{20\sqrt{x}+4x}{3-\sqrt{x}} - \frac{2x^2}{(5\sqrt{x}-x)(3-\sqrt{x})}
$$

Now, it seems, the general parenthesis (and the Eulerian way to designate the unknowns) had become staple food in basic algebra (provided Rotherham is faithful to his manuscript).

A full account is given in [Høyrup, forthcoming].

Coda

This was thus the story of the often meandering transformation of *al-jabr* as received in Latin Europe into the nascent modern algebra of the 17th and 18th centuries, told through the changes of a set of distinctive characteristics.

Let it be said in conclusion that real history was much more meandering. In real history, the characteristics that were traced were not separate; that sometimes shines though the story as I told it, but in the interest of my project I separated as much as I could. Even apart from that, real history was much more winding and bending than illustrated here: after all, I have concentrated on those aspects of the story which either seem to point to the eventual outcome or look as false trails that after all did not contribute to the synthesis as *we* might have expected them to do. As von Ranke (see note 44), I tell "only [partial] histories, not History"

Let is also be said that my story, with all its restrictions and shortcomings, was a story about elements which in synthesis went into the *creation* of the new algebra; it says nothing about how this new tool was *used* by Viète, Descartes and all those who followed; others have done that much better than I could do. Nor does it take up how the "new" algebra itself, originating as a technique for formulating and solving equations, was transformed into *theory;* that other story had started with Cardano, and accelerated in the 17th century. Even that, other friends and colleagues have done better than I would be able to. Mis-paraphrasing Wittgenstein: Where others can speak much better, there one should shut up.

Bibliography

Abdeljaouad, Mahdi, 2005. "Le manuscrit mathématique de Jerba: Une pratique des symboles algébriques maghrébins en pleine maturité", pp. 9–98 *in Actes du 7ème Colloque Maghrébin sur l'Histoire des Mathématiques Arabes*, Marrakech, 30 mai – 1 juin 2002, vol. II. Marrakech: École Normale Supérieure Marrakech.

Abdeljaouad, Mahdi, 2011. "Şeker-Zāde (m. 1787) : Un témoin tardif faisant d'utilisation des symboles mathématiques maghrébins inventés au 12e siècle", pp. 7–32 *in Actes du 10ième Colloque Maghrébin sur l'Histoire des Mathématiques Arabes* (Tunis, 29–30–31 mai 2010). Tunis: Publications de l'Association Tunisienne des Sciences Mathématiques.

Abdeljaouad, Mahdi, 2011. "La circulation des symboles mathématiques maghrébins entre l'Ouest e l'Est musulmans". 9e Colloque maghrébin sur l'histoire des mathématiques arabes (Tipaza, 12–13–14 mai 2007). pp. 7–39 in Abdelmalek Bouzari & Youcef Guergour (eds), *IXème Colloque Maghrebin sur l'histoire des mathématiques arabes*. Tipaza, 12–14 mai 2007 Alger: École Normale Supérieure.

Adam, Charles, & Paul Tannery (eds), 1897. *Oeuvres* de Descartes. I. *Correpondance Avril 1622–Février 1638*. Paris: Léopold Cerf

Adam, Charles, & Paul Tannery (eds), 1902. *Oeuvres* de Descartes. VI. *Discours de la méthode & Essais*. Paris: Léopold Cerf.

Adam, Charles, & Paul Tannery (eds), 1908. *Oeuvres* de Descartes. X. *Physico-mathematica*. Paris: Léopold Cerf.

Alexander, Andreas, 1504. *Mathemalogium prime partis super novam et veterem loycam Aristotelis*. Leipzig: Melchior Lotter.

Arrighi, Gino (ed.), 1967a. Antonio de' Mazzinghi, *Trattato di Fioretti* nella trascelta a cura di M° Benedetto secondo la lezione del Codice L.IV.21 (sec. XV) della Biblioteca degl'Intronati di Siena. Siena: Domus Galilaeana.

Arrighi, Gino, 1967. "Una trascelta delle «miracholose ragioni» di M° Giovanni di Bartolo (secc. XIV–XV)". *Periodico di Matematiche*, 4. Ser. **45**, 11–24.

Arrighi, Gino (ed.), 1987. Paolo Gherardi, *Opera mathematica: Libro di ragioni – Liber habaci*. Codici Magliabechiani Classe XI, nn. 87 e 88 (sec. XIV) della Biblioteca Nazionale di Firenze. Lucca: Pacini-Fazzi.

Arrighi, Gino, 2004/1967. "Nuovi contributi per la storia della matematica in Firenze nell'età di mezzo: Il codice Palatino 573 della Biblioteca Nazionale di Firenze", pp. 159–194 *in* Gino Arrighi, *La matematica nell'Età di Mezzo. Scritti scelti*. Edizioni ETS. First published in *Istituto Lombardo. Accademia di scienze e lettere. Rendiconti, Classe di scienze (A)* **101** (1967), 395–437.

Beeckman, Isaac, 1637. *Catalogus Variorum & insignium Librorum Clarissimi Doctissimique viri D. Isaaci Catalogus ... Quorum Auctio habebitur in aedibus defuncti ad diem 14 Iulij M DC XXXVII*. Dordrecht: Isaac Andreae.

Beeley, Philip, & Christoph J. Scriba (eds), 2005. *The Correspondence of John Wallis*. II. *(1660–September 1668)*. Oxford & New York: Oxford University Press.

Billingsley, Henry (trans.), 1570. *The Elements of Geometrie of the most auncient Philosopher Euclide of Megara*. London: John Daye.

Björnbo, Axel Anthon, 1905. "Gerhard von Cremonas Übersetzung von Alkhwarizmis Algebra und von Euklids Elementen". *Bibliotheca Mathematica*, 3. Folge **6** (1905–06), 239–248.

Bombelli, Rafael, 1579. *L'Algebra*. Bologna: Giovanni Rossi.

Boncompagni, Baldassare, 1851. "Della vita e delle opere di Gherardo cremonese, traduttore des secolo duodecimo, e di Gherardo da Sabbionetta astronomo del secolo decimoterzo". *Atti dell'Accademia pontificia de' Nuovi Lincei* **4** (1850–51), 387–493.

Boncompagni, Baldassare (ed.), 1857. *Scritti* di Leonardo Pisano matematico del secolo decimoterzo. I. Il *Liber abbaci* di Leonardo Pisano. Roma: Tipografia delle Scienze Matematiche e Fisiche.

Boncompagni, Baldassare (ed.), 1862. *Scritti* di Leonardo Pisano matematico del secolo decimoterzo. II. *Practica geometriae* ed *Opusculi*. Roma: Tipografia delle Scienze Matematiche e Fisiche.

Booß-Bavnbek, Bernhelm, & Jens Høyrup, 2021. "Fachwissenschaften und Leitung der Gesellschaft". *Forum Wissenschaft* **38**:3, 55–58.

Bortolotti, Ettore, 1929. *L'algebra, opera di Rafael Bombelli da Bologna. Libri IV e V*. Bologna: Zanichelli.

Bos, Henk J. M., 2001. *Redefining Geometrical Exactness: Descartes' Transformation of the Early Modern Concept of Construction*. New York etc.: Springer.

Briquet, Charles-Moïse, 1923. *Les filigranes. Dictionnaire historique des marques du papier dès leur apparition vers 1282 jusqu'en 1600*. 4 vols. Leipzig: Hiersemann.

Bousquet, Marc-Michel (ed.), 1745. Gottfried Wilhelm Leibniz & Johann Bernoulli, *Commercium philosophicum et mathematicum*. 2 vols. Lausanne: Marc-Michel Bousquet.

Brucker, Gene A., 1969. *Renaissance Florence*. New York etc.: Wiley.

Busard, H. L. L., 1968. "L'algèbre au moyen âge: Le «Liber mensurationum» d'Abû Bekr". *Journal des Savants*, 1968 no. 2, 65–125.

Busard, Hubert L. L., 1976. "Viète, François", pp. 18–25 *in Dictionary of Scientific Biography*, vol. XIV. New York: Scribner.

Buteo, Joannes, 1559. *Logistica, quae et arithmética vulgò dicitur in libros quinque digesta*. Lyon: Guillaume Rouillé.

Cajori, Florian, 1928. *A History of Mathematical Notations*. I. *Notations in Elementary mathematics*. II. *Notations Mainly in Higher Mathematics*. La Salle, Illinois: Open Court, 1928–29.

Calzoni, Giuseppe, & Gianfranco Cavazzoni (eds), 1996. Luca Pacioli, *"tractatus mathematicus ad discipulos perusinos"*. Città di Castello: Delta Grafica.

Cardano, Girolamo, 1539. *Practica arithmetice, et mensurandi singularis*. Milano: Bernardini Calusco.

Cardano, Girolamo, 1545. *Artis magnae sive de regulis algebraicis, liber unus*. Nürnberg: Johan Petreius.

Cardano, Girolamo, 1570. *Opus novum de proportionibus numerorum, motuum* [...] *Praeterea artis magnae sive de regulis algebraicis liber unus. Item de aliza regula liber*. Basel: Henricpetrina.

Cardano, Girolamo, 1663. Hieronymo Cardani Mediolanensis Philosophi ac Medici celeberrimi *Operum* tomus quartus; quo continentur *Arithmetica, Geometrica, Musica*. Lyon: Jean Antoine Huguetan & Marc Antoine Ragaud.

Cassinet, Jean, 2001. "Une arithmétique toscane en 1334 en Avignon dans la citè des papes et de leurs banquiers florentins", pp. 105–128 *in Commerce et mathématiques du moyen âge à la renaissance, autour de la Méditerranée*. Toulouse: Éditions du C.I.H.S.O.

Clavius, Cristopher, 1608. *Algebra*. Roma: Bartolomeo Zanetti.

Confalonieri, Sara, 2013. "The Telling of the Unattainable Attempt to Avoid the *casus irreducibilis* for Cubic Equations: Cardano's *De Regula Aliza*". *Thèse de doctorat*, Université Paris Diderot.

Corominas, Joan, & José A. Pascual, 1980. *Diccionario Crítico etimológico castellano a hispánico.* 6 vols. Madrid: Gredos, 1980–1983.

Costa, Mercè, & Maribel Tarrés, 2001. *Diccionari del català antic*. Barcelona: Edicions 62.

Curtze, Maximilian (ed.), 1902. *Urkunden zur Geschichte der Mathematik im Mittelalter und der Renaissance.* Leipzig: Teubner.

de Crouzas, Jean-Pierre, 1726. *Traite de l'algebre*. Paris: François Montalant.

de la Roche, Etienne, 1520. *Larismethique novellement composee.* Lyon: Constantin Fradin.

de l'Hospital, Guillaume François Antoine, 1696. *Analyse des infiniment petits, pour l'intelligence des lignes courbes.* Paris: Imprimerie Royale

Di Teodoro, Francesco Paolo, 2014. "Pacioli, Luca". *Dizionario Biografico degli Italiani* **80** (2014), online. https://www.treccani.it/enciclopedia/luca-pacioli_%28Dizionario-Biografico%29/ (accessed 18.9.2021).

Delle Donne, Fulvio, 2007. "Un'inedita epistola sulla morte di Guglielmo de Luna, Maestro presso lo studium di Napoli, e le traduzione prodotte alla corte di Manfredi di Svevia". *Recherches de Théologie et Philosophie médiévales* **74**, 225–245.

Descartes, René, 1637. *Discours de la methode pour bien conduire sa raison, & chercher la verité dans les sciences. Plus La dioptrique. Les meteores. Et La geometrie.* Leiden: Ian Maire.

Eamon, William, 1991. "Court, Academy and Printing House: Patronage and Scientific Careers in Late Renaissance Italy", pp. 25–50 *in* Bruce T. Moran (ed.), *Patronage and Institutions. Science, Technology and Medicine at the European Court, 1500–1750.* Rochester, New York: Boydell & Brewer.

Euler, Leonhard, 1748. *Introductio in analysin infinitorum.* 2 vols. Lausanne: Bousquet.

Federici Vescovini, Graziella, 1968. "Bianchini, Giovanni". *Dizionario Biografico degli Italiani* **10** (1968). https://www.treccani.it/enciclopedia/giovanni-bianchini_(Dizionario-Biografico) (accessed 8.10.2021).

Franci, Raffaella, & Marisa Pancanti (eds), 1988. Anonimo (sec. XIV), *Il trattato d'algibra* dal manoscritto Fond. Prin. II. V. 152 della Biblioteca Nazionale di Firenze. Siena: Servizio Editoriale dell'Università di Siena

Franci, Raffaella (ed.), 2001. Maestro Dardi, *Aliabraa argibra*, dal manoscritto I.VII.17 della Biblioteca Comunale di Siena. Siena: Università degli Studi di Siena.

Franci, Raffaella, 2021. "Una volgarizzazione quattrocentesca dell'algebra di al-Khwārizmī e Guglielmo de Lunis traduttore dell *al-jabr*". *Bollettino di storia delle scienze matematiche* **41**, 231–261.

Gärtner, Barbara (ed.), 2000. *Johannes Widmanns »Behende vnd hubsche Rechenung«. Die Textsorte »Rechenbuch« in der frühen Neuzeit.* Tübingen: Max Niemeyer.

Ghaligai, Francesco, 1521. *Summa de arithmetica.* Firenze: Bernardo Zucchetta.

Giusti, Enrico, 2017. "The Twelfth Chapter of Fibonacci's *Liber abbaci* in Its 1202 Version". *Bollettino di Storia delle Scienze Matematiche* **37**, 9–216.

Giusti, Enrico (ed.), 2020. Leonardi Bigolli Pisano *vulgo* Fibonacci *Liber Abbaci*. Firenze: Leo S. Olschki.

Gosselin, Guillaume, 1577. *De arte magna, seu de occulta parte numerorum quae et Almucabala vulgo dicitur.* Paris: Egide Beys.

Gosselin, Giullaume (ed., trans.), 1578. *L'arithmetique* de Nicolas Tartaglia brescian, grand mathematicien, et prince des praticiens. Divisée en deux parties. Paris: Gilles Beys.

Gragg, Florence Alden (ed.), 1927. *Latin Writings of the Italian Humanists; Selections*. New York: Scribner.

Grammateus, Heinrich, 1521. *Ayn new kunstlich Buech, welches gar gewiß und behend lernet nach der gemainen regel Detre, welschen Practic, regel falsi unn etlichen regeln Cosse*. Wien: Lucas Alantsee.

Gregori, Silvano, & Lucia Grugnetti (eds), 1998. Anonimo (sec. XV), *Libro di conti e mercatanzie*. Parma: Università degli Studi di Parma, Dipartimento di Matematica.

Hammond, Nathaniel, 1742. *The Elements of Algebra in a New and Easy Method*. Cornhill: J. Walthor.

Harriot, Thomas, 1631. *Artis analyticae praxis*. London: Robert Barker.

Heath, Thomas L. (ed., trans.), 1926. *The Thirteen Books of Euclid's Elements*. 3 vols. Cambridge: Cambridge University Press / New York: Macmillan.

Heeffer, Albrecht, 2012. "The Rule of Quantity by Chuquet and de la Roche and its Influence on German Cossic Algebra", pp. 127–147 *in* Sabine Rommevaux, Maryvonne Spiesser & Maria Rosa Massa Esteve (eds), 2012. *Pluralité de l'algèbre à la Renaissance*. Paris: Honoré Champion.

Heeffer, Albrecht, 2015. Draft transcription of Bianchini, *Flores Almagesti*, algebra section.

Heiberg, Johan Ludvig (ed., trans.), 1912. Heronis *Definitiones* cum variis collectionibus. Heronis quae feruntur *Geometrica*. Leipzig: Teubner.

Hellmann, Martin, 2003. "Von Pythagoras zum Liber de cosa: Bemerkungen von Nicolaus Matz über die Geschichte der Rechenkunst", pp. 75–112 *in* Wolfgang Schmitz (ed.), *Bewahren und Erforschen*. Festgabe für Kurt Hans Staub zum 70. Geburtstag. Michelstadt: Stadt Michelstadt.

Hochheim, Adolf (ed., trans.), 1878. *Kâfî fîl Hisâb (Genügendes über Arithmetik) des Abu Bekr Muhammed ben Alhusein Alkarkhi*. I-III. Halle: Louis Nebert, 1878–1880.

Høyrup, Jens, 1988. "Jordanus de Nemore, 13th Century Mathematical Innovator: an Essay on Intellectual Context, Achievement, and Failure". *Archive for History of Exact Sciences* **38** (1988), 307–363.

Høyrup, Jens, 1998. "'Oxford' and 'Gherardo da Cremona': on the Relation between Two Versions of al-Khwārizmī's Algebra", pp. 159–178 *in Actes du 3me Colloque Maghrébin sur l'Histoire des Mathématiques Arabes, Tipaza (Alger, Algérie), 1–3 Décembre 1990*, vol. II. Alger: Association Algérienne d'Histoire des Mathématiques, 1998. Reprint without the copious typesetting errors in [Høyrup 2019a].^R

Høyrup, Jens, 2002. "Pedro Nuñez: Innovateur bloqué, et dernier témoin d'une tradition millénaire". *Gazeta de Matemática* n° 143, 52–59.

Høyrup, Jens, 2007. *Jacopo da Firenze's Tractatus Algorismi and Early Italian Abbacus Culture*. Basel etc.: Birkhäuser.

Høyrup, Jens, 2009. "What Did the Abbacus Teachers Aim At When They (Sometimes) Ended Up Doing Mathematics? An Investigation of the Incentives and Norms of a Distinct Mathematical Practice", pp. 47–75 *in* Bart van Kerkhove (ed.), *New Perspectives on Mathematical Practices: Essays in Philosophy and History of Mathematics*. Singapore: World Scientific.

Høyrup, Jens, 2010. "Hesitating progress – the slow development toward algebraic symbolization in abbacus- and related manuscripts, c. 1300 to c. 1550", pp. 3–56 *in* Albrecht Heeffer & Maarten Van Dyck (eds), *Philosophical Aspects of Symbolic Reasoning in Early Modern Mathematics*. London: College Publications.

Høyrup, Jens, 2011. "A Diluted al-Karajī in Abbacus Mathematics", pp. 187–197 *in Actes du 10ième Colloque Maghrébin sur l'Histoire des Mathématiques Arabes* (Tunis, 29–30–31 mai 2010). Tunis: Publications de l'Association Tunisienne des Sciences Mathématiques.

Høyrup, Jens, 2015. "Embedding: Another Case of stumbling progress". *Physis* **50**, 1–38.

Høyrup, Jens, 2019a. *Selected Essays on Pre- and Early Modern Mathematical Practice*. Cham etc.: Springer.

Høyrup, Jens, 2019b. "Reinventing or Borrowing Hot Water? Early Latin and Tuscan Algebraic Operations with Two Unknowns". *Ganita Bhāratī* **41**, 115-159.

Høyrup, Jens, 2020. "Fifteenth-Century Italian Symbolic Algebraic Calculation with Four or Five Unknowns". *Ganita Bhāratī* **42**, 161–192.

Høyrup, Jens, 2021. "Advanced Arithmetic from Twelfth-Century Al-Andalus, Surviving Only (and Anonymously) in Latin Translation? A Narrative That Was Never Told", pp. 33–61 *in* Sonja Brentjes and Alexander Fidora (eds), *Premodern Translation: Comparative Approaches to Cross-Cultural Transformations*. Turnhout: Brepols.

Høyrup, Jens, 2021b. "Peeping into Fibonacci's Study Room". *Ganita Bhāratī* **43**, 1–70.

Høyrup, Jens, 2022. "Intermediaries between Abū Kāmil's and Fibonacci's Algebras – Lost but Leaving Indubitable Traces", pp. 173–189 *in* Hmida Hedfi & Mahdi Abdeljaouad (eds), *Actes du XIVe Colloque Maghrébin sur l'Histoire des Mathmatiques Arabes*, français-anglais. Tunis.

Høyrup, Jens, 2024. *The World of the Abbaco: Abbacus Mathematics Analyzed and Situated Historically between Fibonacci and Stifel*. Cham: Birkhäuer, 2024.

Høyrup, Jens, forthcoming. "How did the all-purpose parenthesis come about in European algebra?". To appear in *Ganita Bhāratī*.

Hughes, Barnabas (ed.), 1986. "Gerard of Cremona's Translation of al-Khwārizmī's *Al-Jabr*: A Critical Edition". *Mediaeval Studies* **48**, 211–263.

Itard, Jean, 1971. "Chuquet, Nicolas", pp. 272–278 *in Dictionary of Scientific Biography* III. New York: Scribner.

l'Huillier, Ghislaine (ed.), 1990. Jean de Murs, le *Quadripartitum numerorum*. Genève & Paris: Droz.

Lamrabet, Driss, 1994. *Introduction à l'histoire des mathématiques maghrébines*. Rabat: The Author.

Libri, Guillaume, 1838. *Histoire des mathématiques en Italie*. 4 vols. Paris: Jules Renouard, 1838–1841.

Lindemann, Ferdinand, 1904. "Lehren und Lernen der Mathematik". Rede beim Antritt des Rektorats der Ludwig-Maximiliams-Universität gehalten am 26. November 1904. München: C. Wolf & Sohn.

Loget, François, 2012. "L'algèbre en France au XVIe siècle: individus et réseax", pp. 69–101 *in* Sabine Rommevaux, Maryvonne Spiesser & Maria Rosa Massa Esteve (eds), 2012. *Pluralité de l'algèbre à la Renaissance*. Paris: Honoré Champion.

Luckey, Paul, 1941. "Tābit b. Qurra über den geometrischen Richtigkeitsnachweis der Auflösung der quadratischen Gleichungen". *Sächsischen Akademie der Wissenschaften zu Leipzig. Mathematisch-physische Klasse. Berichte* **93**, 93–114.

Maia Bertato, Fábio (ed.), 2008. Luca Pacioli, *De divina proportione*. Tradução Anotada e Comentada. Versão Final da Tese (Doutorado em filosofia – CLE/IFCH/UNICAMP). Campinas.

Marre, Aristide (ed.), 1881. *Le Triparty en la science des nombres* par Maistre Nicolas Chuquet Parisien, précédé d'une notice. Roma: Imprimerie des Sciences Mathématiques et Physiques.

Mennher, Valentin, 1556. *Arithmetique seconde*. Antwerpen: Ian Loë.

Mennher, Valentin, 1564. *Praticque des triangles sphériques, des distances sur les globes, et autres ingenieuses et nouvelles questions mathematique.* Antwerpen: Gilles Coppenius.

Mennher, Valentin, 1565. *Praticque pour brievement apprendre à ciffrer, et tenir livre de compte, avec la regle de coss, et geometrie.* Antwerpen.

Meskens, Ad, 2013. *Practical Mathematics in a Commercial Metropolis: Mathematical Life in Late 16th Century Antwerp.* Dordrecht etc.: Springer.

Miura, Nobuo, 1981. "The Algebra in the *Liber abaci* of Leonardo Pisano". *Historia Scientiarum* **21**, 57–65.

Moyon, Marc, 2019a. "The *Liber Restauracionis:* A Newly Discovered Copy of a Mediaeval Algebra in Florence". *Historia Mathematica* **46**, 1–37.

Moyon, Marc, 2019b. "Le *Liber augmenti et diminutionis:* Contribution à l'histoire des mathématiques médiévales". *Mémoire, Habilitation à diriger des recherches*, Université de Limoges.

Nefe, Mattheus, 1565. *Arithmetica: Zwey newe Rechenbuecher.* Breslau.

Nesselmann, G. H. F., 1842. *Versuch einer kritischen Geschichte der Algebra.* Erster Theil, *Die Algebra der Griechen.* Berlin: G. Reimer.

Newton, Isaac, 1707. *Arithmetica universalis; sive, de compositione resolutione arithmetica liber.* Cambridge: Typis Academicis / London: Benj. Tooke, 1707.

Newton, Isaac, 1711. *Analysis per quantitatum series, fluxiones, ac differentias.* London: Pearson.

Nuñez, Pedro, 1567. *Libro de Algebra en Arithmetica y Geometria.* Antwerpen: Arnaldo Birckman.

Pacioli, Luca, 1494. *Summa de Arithmetica Geometria Proportioni et Proportionalita.* Venezia: Paganino de Paganini.

Pacioli, Luca, 1509. *Divina proportione.* Venezia: Paganius Paganinus.

Parkes, M. B., 2016. *Pause and Effect: An Introduction to the History of Punctuation in the West.* London & New York: Routledge.

Peletier, Jacques, 1554. *L'algebre.* Lyon: Ian de Tournes.

Peletier, Jacques, 1560. *De occulta parte numerorum, quam algebram vocant, libri duo.* Paris: Guillaume Cavellat.

Peucer, Caspar, 1556. *Logistice Astronomica Hexacontadôn* Item: *Logistice Regulae Arithmeticae, qvam Cossam & Algebram quadratam vocant.* Wittenberg: Georg Rhau.

Pieraccini, Lucia (ed.), 1983. M° Biagio, *Chasi exenplari alla regola dell'algibra nella trascelta a cura di M° Benedetto* dal Codice L. VII. 2Q della Biblioteca Comunale di Siena. Siena: Servizio Editoriale dell'Università di Siena.

Procissi, Angiolo (ed.), 1954. "I Ragionamenti d'Algebra di R. Canacci". *Bollettino Unione Matematica Italiana*, serie III, **9**, 300–326, 420–451.

[Ramus], Petrus, 1560. *Algebra.* Paris: Andreas Wechelum.

Rashdall, Hastings, 1936. *The Universities of Europe in the Middle Ages.* Oxford: The Clarendon Press.

Rashed, Roshdi, & Ahmed Djebbar (ed., trans.), 1981. *L'oeuvre algébrique d'al-Khayyām.* Aleppo: University of Aleppo, Institute for the History of Arabic Science.

Rashed, Roshdi (ed., trans.), 2007. Al-Khwārizmī, *Le Commencement de l'algèbre.* Paris: Blanchard.

Rashed, Roshdi (ed., trans.), 2012. Abū Kāmil, *Algèbre et analyse Diophantienne.* Berlin & Boston: de Gruyter.

Raynouard, François-Just-Marie, 1838. *Lexique roman, ou, Dictionnaire de la langue des troubadours.* 6 vols. Paris: Silvestre, 1838–1844.

Ristori, Renzo, 1979. "Cegia, Francesco", in *Dizionario biografico degli italiani*, vol. 23. Roma: Instituto della Enciclopedia Italiana. https://www.treccani.it/enciclopedia/francesco-cegia_(Dizionario-Biografico)/ (accessed 5.4.2021).

Rotherham, W., 1852. *The Algebraical Equation and Problem Papers, Proposed in the Examinations of St. John's College, Cambridge, from the Year 1794 to the Present Time*. Cambridge: Henry Wallis.

Rüdiger, Bernd, Rainer Gebhardt & Menso Folkerts (eds), 2023. Adam Ries, *Coß 1*. 2 vols. Annaberg-Buchholz: Adam-Ries-Bund.

Rudolff, Christoff, 1525. *Behend unnd hübsch Rechnung durch die kunstreichen Regeln Algebra, so gemeincklich die Coss genennt werden*. Straßburg.

Salomone, Lucia (ed.), 1982. M° Benedetto da Firenze, *La reghola de algebra amuchabale* dal Codice L.IV.21 della Biblioteca Comunale de Siena. Siena: Servizio Editoriale dell'Università di Siena.

Sammarchi, Eleonora, 2019. "Les collections de problèmes algèbriques dans le *Qistās al-muʿādala fī ʾilm al-jabr waʾl-muqābala* d'al-Zanjānī". *Méviévales* **66**, 37–56.

Sangster, Alan, 2007. "The Printing of Pacioli's Summa in 1494: How Many Copies Were Printed?" *Accounting Historians Journal* **34**, 125-145.

Scheubel, Johann (ed., trans.), 1550. Euclidis Megarensis, Philosophi et mathematici excellentissimi, *Sex libri priores*. Basel: Herwagen.

Scheubel, Johann, 1551. *Algebrae compendiosa facilísque descriptio, qua depromuntur magna Arithmetices miracula*. Parisiis: Guillaume Cavellat.

Sesiano, Jacques, 2000. "Un recueil du XIIIᵉ siècle de problèmes mathématiques". *SCIAMUS* **1**, 71-132.

Sesiano, Jacques, 2014. *The* Liber mahameleth: *A Twelfth-Century Mathematical Treatise*. Heidelberg etc.: Springer.

Simi, Annalisa (ed.), 1994. Anonimo (sec. XIV), *Trattato dell'alcibra amuchabile* dal Codice Ricc. 2263 della Biblioteca Riccardiana di Firenze. Siena: Servizio Editoriale dell' Università di Siena.

Steinschneider, Moritz, 1904. "Die europäischen Übersetzungen aus dem Arabischen bis Mitte des 17. Jahrhunderts". I-II. *Sitzungsberichte der Kaiserlichen Akademie der Wissenschaften in Wien, philosophisch-historische Klasse* CXLIX/iv (1904), CLI/i (1905).

Stevin, Simon, 1585a. *L'aritmetique*. Leiden: Plantin.

Stevin, Simon, 1585b. *De thiende*. Leiden: Christoffel Plantin, 1585. Facsimile pp. 386–424 *in* Dirk J. Struik (ed.), *The Principal Works of Simon Stevin*. Vol. IIA–B, *Mathematics*. Amsterdam: C. V. Swets & Zeitlinger, 1958.

Stifel, Michael, 1544. *Arithmetica integra*. Nürnberg: Johan Petreius.

Stifel, Michael (ed.), 1553. *Die Coss* Christoffs Rudolffs. Die schönen Exemplen der Coss gebessert und gemehrt. Königsberg in Preussen: Alexander Lutomyslensis.

Suter, Heinrich, 1910. "Das Buch der Seltenheiten der Rechenkunst von Abū Kāmil al-Miṣrī". *Bibliotheca Mathematica*, 3. Folge **11** (1910–1911), 100–120.

Tartaglia, Nicolò, 1546. *Quesiti et inventioni diverse*. Venezia: Venturino Ruffinelli.

Tartaglia, Nicolò, 1556. *La seconda parte del general trattato di numeri, et misure*. Venezia: Curtio Troiano.

Tartaglia, Nicolò, 1560. *La sesta parte del general trattato de' numeri, et misure*. Venezia: Curtio Troiano.

Ulivi, Elisabetta, 1996. "Per una biografia di Antonio Mazzinghi, maestro d'abaco del XIV secolo". *Bollettino di Storia delle Scienze Matematiche* **16**, 101-150.

Ulivi, Elisabetta, 2002. "Benedetto da Firenze (1429–1479), un maestro d'abaco del XV secolo. Con documenti inediti e con un'Appendice su abacisti e scuole d'abaco a Firenze nei secoli XIII–XVI". *Bollettino di Storia delle Scienze Matematiche* **22**, 3–243.

Ulivi, Elisabetta, 2004. "Maestri e scuole d'abaco a Firenze: La Bottega di Santa trinità". *Bollettino di storia delle scienze matematiche* **24**, 43–91.

Ulivi, Elisabetta, 2017. *Giovanni del Sodo, un maestro d'abaco fiorentino nel* Tractatus mathematicus ad discipulos perusinos *di Luca Pacioli*. Firenze: Università di Firenze, Dipartimento di Matematica e Informatica.

Valla, Giorgio, 1501. *De expetendis et fugiendis rebus opus*. 2 vols. Venezia: Manutius.

van Randenborgh, Christian, 2012. "Frans van Schootens Beitrag zu Descartes Discours de la méthode". *Mathematische Semesterberichte* **59**, 233–241.

van Schooten, Frans (ed.), 1646. François Viète, *Opera mathematica*. Leiden: Elzevier.

Van Egmond, Warren, 1978. "The Earliest Vernacular Treatment of Algebra: The *Libro di ragioni* of Paolo Gerardi (1328)". *Physis* **20**, 155–189.

Van Egmond, Warren, 1980. *Practical Mathematics in the Italian Renaissance: A Catalog of Italian Abbacus Manuscripts and Printed Books to 1600*. Firenze: Istituto e Museo di Storia della Scienza.

Van Egmond, Warren, 1983. "The Algebra of Master Dardi of Pisa". *Historia Mathematica* **10**, 399–421.

Van Egmond, Warren (ed.), 1986. Anonimo (sec. XV), *Della radice de' numeri e metodo di trovarla (Trattatello di Algebra e Geometria) dal Codice Ital. 578 della Biblioteca Estense di Modena. Parte prima.* Siena: Servizio Editoriale dell'Università di Siena.

Vasset, Antoine, 1630. *L'algebre nouvelle de M'. Viète*, traduite en françois. Paris: Pierre Rocolet.

Vaulezard, Jean-Louis, 1630. *Les cinq livres des zetetiques de Francois Viette*. Paris: Julian Jacquin.

Viète, François, 1591a. *In artem analyticem isagoge*. Tours: Jamet Mettayer.

Viète, François, 1591b. *Zeteticorum libri quinque*. Tours: Jamet Mettayer.

Viète, François, 1593a. *Variorum de rebus mathematicis responsorum, liber VIII*. Tours: Jamet Mettayer.

Viète, François, 1593b. *Supplementum geometriae: Ex opera restitutae Mathematicae Analyseos, seu Algebrâ novâ*. Tours: Jamet Mettayer.

Vogel, Kurt, 1971. "Fibonacci, Leonardo", pp. 604–613 *in Dictionary of Scientific Biography* IV. New York: Scribner.

Vogel, Kurt, 1977. *Ein italienisches Rechenbuch aus dem 14. Jahrhundert (Columbia X 511 A13)*. München.

Vogel, Kurt (ed.), 1981. *Die erste deutsche Algebra aus dem Jahre 1481*. München: Verlag der Bayerischen Akademie der Wissenschaften.

von Ranke, Leopold, 1824. *Zur Kritik neuerer Geschichtsschreiber. Eine Beylage zu desselben romanischen und germanischen Geschichten*. Leipzig & Berlin: Reimer.

Wallis, John, 1655. *De sectionibus conicis nova methodo expositis tractatus*. Oxford: Thomas Robinson.

Wallis, John, 1658. *Commercium epistolicum de Quaestionibus quibusdam Mathematicis nuper habitum*. Oxford: Thomas Robinson.

Wallis, John, 1685. *A Treatise of Algebra, both Historical and Practical*. Oxford: Richard Davis.

Wappler, Hermann Emil, 1887. "Zur Geschichte der deutschen Algebra im 15. Jahrhundert", pp. 1–32 *in Gymnasium zu Zwickau. Jahresbericht über das Schuljahr von Ostern 1886 bis Ostern*

1887. Zwickau: R. Zückler.

Westfall, Richard S., 1980. *Never at Rest: A Biography of Isaac Newton*. Cambridge: Cambridge University Press.

Witmer, T. Richard (ed., trans.), 1983. François Viète, *The Analytic Art: Nine Studies in Algebra, Geometry, and Trigonometry from the Opus restitutae mathematicae analyseos, seu algebrâ novâ*. Kent, Ohio: Kent State University Press.

Woepcke, Franz, 1853. *Extrait du Fakhrî, traité d'algèbre* par Aboû Bekr Mohammed ben Alhaçan Alkarkhî. Paris: L'Imprimerie Impériale.

Woepcke, Franz, 1854. "Recherches sur l'histoire des sciences mathématiques chez les Orientaux, d'après des traités inédits arabes et persans. Premier article. Notice sur des notations algébriques employées par les Arabes". *Journal Asiatique*, 5e série **4**, 348–384.

Zamberti, Bartolomeo (ed., trans.), 1505. *Euclidis megarensis philosophi platonici, Mathematicarum disciplinarum janitoris, habent in hoc volumine quicunque ad mathematicam substantiam aspirant: Elementorum libros xiii cum expositione insignis mathematici ...* Venezia: Giovanni Tacuini.

Youschkevitch, Adolf P., 1976. "The Concept of Function up to the Middle of the 19th Century". *Archive for History of Exact Sciences* **16**, 37–85

Xylander, Wilhelm (ed., trans.), 1575. Diophanti Alexandrini *Rerum arithmeticarum libri sex*. Basel: Eusebius Episcopius.

Index